CORROSION OF REINFORCING STEEL IN CONCRETE

A symposium
sponsored by ASTM
Committee G-1 on
Corrosion of Metals
AMERICAN SOCIETY FOR
TESTING AND MATERIALS
Bal Harbour, Fla., 4–5 Dec. 1978

ASTM SPECIAL TECHNICAL PUBLICATIONS 713
D. E. Tonini, American Hot Dip Galvanizers
Association, Inc., and J. M. Gaidis,
W. R. Grace & Co., editors

04-713000-27

AMERICAN SOCIETY FOR TESTING AND MATERIALS
1916 Race Street, Philadelphia, Pa. 19103

Library of Congress Catalog Card Number: 80-65020

NOTE

The Society is not responsible, as a body,
for the statements and opinions
advanced in this publication.

Printed in Baltimore, Md.
August 1980

Second Printing, Baltimore, Md.
July 1984

Foreword

The symposium on Corrosion of Reinforcing Steel in Concrete was presented at Bal Harbour, Fla., 4–5 Dec. 1978. The American Society for Testing and Materials' Committee G-1 on Corrosion of Metals through its Subcommittee G01.14 on Reinforcing Steel in Concrete sponsored the symposium. D. E. Tonini, Albert Cook, and H. M. Maxwell presided as symposium cochairman. D. E. Tonini, American Hot Dip Galvanizers Association, Inc., and J. M. Gaidis, W. R. Grace & Co., are editors of this publication.

Related ASTM Publications

Stress Relaxation Testing, STP 676 (1979), 04-676000-23

Stress Corrosion Cracking—The Slow Strain-Rate Technique, STP 665 (1979), 04-665000-27

Corrosion Fatigue Technology, STP 642 (1978), 04-642000-27

Chloride Corrosion of Steel in Concrete, STP 629 (1977), 04-629000-27

Corrosion, Wear, and Interfaces with Corrosion, STP 567 (1974), 04-567000-29

Corrosion in Natural Environments, STP 558 (1974), 04-558000-27

A Note of Appreciation to Reviewers

This publication is made possible by the authors and, also, the unheralded efforts of the reviewers. This body of technical experts whose dedication, sacrifice of time and effort, and collective wisdom in reviewing the papers must be acknowledged. The quality level of ASTM publications is a direct function of their respected opinions. On behalf of ASTM we acknowledge with appreciation their contribution.

ASTM Committee on Publications

Editorial Staff

Contents

Introduction

The effects of corrosion of reinforcing steel in concrete subjected to chloride environments have been observed for at least 50 years. However, efforts to quantify the corrosion mechanisms involved have largely been confined to the past decade. These efforts have been stimulated significantly by the problems created by the increase in deicing salt usage on the U.S. Interstate Highway system. The costs associated with this corrosion are known to be heavy, although their true magnitude remains a matter for discussion.

As an outgrowth of the interest generated by the highway bridge situation, ASTM Subcommittee G01.04 organized a symposium on "Chloride Corrosion of Steel in Concrete," which was presented at the 79th Annual Meeting of the American Society for Testing and Materials in Chicago, Ill. 27 June–2 July 1976. The purpose of the symposium was to bring together the experience of laboratory and field engineers who had dealt with this problem. Papers presented at the conference and later published as *ASTM STP 629, Chloride Corrosion of Steel in Concrete* were intended to provide researchers and engineers with a convenient compilation of information and recommendations. This compilation was, in effect, a report on the state of the art with respect to control technology being used during the mid-1970's.

Following the 1976 symposium, ASTM Subcommittee G01.14 on Corrosion of Reinforcing Steel in Concrete was formed to provide an expanded forum for those concerned with testing and materials for coastal or offshore reinforced concrete structures as well as highway bridges. One of the early orders of business for the Subcommittee was to organize its work into three task group efforts: (1) "Test Methods and Monitoring of Corrosion in New and Repaired Concrete Structures"; (2) "The Effect of Electrical Grounding, Galvanic Couples, and Stray Currents on Reinforcement in Concrete"; and (3) "Corrosion Mechanisms and Laboratory Evaluation of Corrosion Resistance of Reinforcement."

As a consequence of the scope reflected in the G01.14 Subcommittee structure, it was decided to organize a second symposium to report additional data, particularly with regard to the testing aspects of corrosion of reinforcing steel in concrete. Thc symposium was held during ASTM Committee Week, 3-8 Dec. 1978, in Bal Harbour, Fla. In contrast to the papers presented in Chicago in 1976, the Bal Harbour papers reflect a generally stronger academic and more rigorous approach to both materials and testing subject matter.

We wish to express our appreciation to the authors and to A. R. Cook, Chairman of G01.14; H. M. Maxwell, member of G01.14 and symposium vice chairman; and C. B. Sanborn, secretary of G01.14, for their invaluable assistance in organizing and presenting the symposium.

D. E. Tonini

American Hot Dip Galvanizers Association, Inc., Washington, D.C. 20005; symposium cochairman and coeditor.

J. M. Gaidis

W. R. Grace & Co., Columbia, Md. 21044; symposium cochairman and coeditor.

C. E. Locke[1] *and A. Siman*[1]

Electrochemistry of Reinforcing Steel in Salt-Contaminated Concrete

REFERENCE: Locke, C. E. and Siman, A., "**Electrochemistry of Reinforcing Steel in Salt-Contaminated Concrete,**" *Corrosion of Reinforcing Steel in Concrete, ASTM STP 713,* D. E. Tonini and J. M. Gaidis, Eds., American Society for Testing and Materials, 1980, pp. 3-16.

ABSTRACT: Corrosion rates of reinforcing steel have been measured in concrete using the polarization resistance technique. The corrosion rates have been calculated for seven different sodium chloride content and two different surface conditions of steel in concrete. The results from partially coated reinforcing steel specimens indicate the existence of a critical sodium chloride concentration between 0.1 and 0.2 percent by weight of concrete at which the rate of corrosion increases significantly. Anodic and cathodic Tafel slopes have also been determined experimentally. The high values of Tafel slopes may be attributed in part to IR drop; however, more research is needed to clarify this matter.

KEY WORDS: reinforcing steel, corrosion, concrete, bridge decks, chlorides, corrosion rate, polarization resistance, Tafel slope

The widespread use of reinforced concrete structures has attracted investigators to study the problems associated with corrosion of steel in concrete. It is generally believed that due to the high alkalinity of concrete environments (pH 12.5) a protective layer is formed on the surface of the steel which provides adequate corrosion resistance. However, small amounts of Cl^- will destroy this inhibitive property of concrete. Reinforced concrete construction exposed to high Cl^- environments, such as marine structures and bridge decks, experiences premature deterioration and failure.

The increased use of salts to remove snow and ice in the past several years in the United States has resulted in severe damage to bridge decks and other reinforced highway structures. The salt usage has increased from less than 0.5 million tons in 1947 to 10 million tons in 1975 [*1*].[2] This consumption rate has resulted in more frequent repair of bridge decks. The estimated an-

[1]Associate professor and graduate student, respectively, University of Oklahoma, School of Chemical Engineering and Materials Science, Norman, Okla. 73019.

[2]The italic numbers in brackets refer to the list of references appended to this paper.

nual cost of bridge deck repairs was $70 million in 1973, which increased to $200 million by 1975 [*1*].

In recent years, extensive efforts have been made to fully investigate the electrochemical behavior of steel in concrete. The corrosion process of steel in concrete is a function of many variables such as the steel surface, concrete properties, and the environment in which the concrete is used. These complicate the study of the phenomenon, including quantitative measurements. Numerous methods have been proposed to prevent or hinder the corrosion problem of bridge decks and some have already been applied. Since any preventive technique has to be tested before and during field application, the improvement of testing techniques is of great importance.

Corrosion rate measurement is a reliable approach which has been used to investigate corrosion processes for many years. Corrosion rates of steel in concrete reported in the literature are generally from weight loss tests, visual inspection, or average pit depth. Raphael and Shalon [*2*], Alekseev and Rozental [*3*], and Akimova [*4*] have reported corrosion rates using weight loss tests at different conditions. Recently some electrical and electrochemical techniques have also been used for determination of corrosion rates of steel in concrete. Griffin and Henry [*5*] have reported relatively high corrosion rates by imbedding electrical resistance probes in concrete. Dawson et al [*6*] investigated corrosion rates by using the a-c impedance method. The polarization resistance technique has been applied successfully for corrosion rate measurement in many industrial environments. This technique has also drawn a lot of attention for application to corrosion of steel in concrete. Gouda et al [*7*] measured corrosion rates by using the polarization resistance technique. Table 1 shows some reported data on corrosion rates using these different techniques. Although these experiments have been conducted at totally different conditions, the results are generally the same order of magnitude. In the present investigation, the polarization resistance technique has been used to determine the corrosion rate of reinforcing steel in salt-contaminated portland cement concrete.

Experimental

The corrosion rate of reinforcing steel at two different surface conditions has been studied. In the first set, three-electrode polarization probes were made from reinforcing steel. One-half-inch (12.7 mm) reinforcing steel bars were machined to 0.63 cm (¹/₄ in.) and three pieces were mounted symmetrically 1.35 cm (0.53 in.) apart in an epoxy resin as shown in Fig. 1. The exposed surface area of each electrode was 9 cm^2 (1.39 $in.^2$). Each probe was then washed with acetone and cast in 15.2 by 30.5 cm (6 by 12 in.) concrete cylinders. These probes are referred to as "machined probes." Seven different batches of concrete were made, each containing a different sodium chloride concentration ranging from 0 to 1 percent (weight percent of total

TABLE 1—*Reported values of corrosion rate.*

Corrosion rate, mils per year	Procedure and Details	Technique	Reference
0.01 to 0.12	no salt exposed to different climatic conditions up to 5 years (RH[a] = 55 to 95% Temperature = 20 to 50°C); W/C[b] = 0.6 kg water/kg cement	weight loss	Raphael and Shalon [2]
0.124 to 0.887	addition of 5%[c] $CaCl_2$, after 3 years' storage at RH = 90%; W/C = 0.5 to 0.7	weight loss	Alekseev and Rozental [3]
0.085 to 0.765	4%[c] NaCl in concrete for 7 months in air or oxygen atmosphere (pressure 2 atm) RH = 75%; W/C = 0.4 and 0.5	weight loss	Akimova [4]
0 to 12 with a maximum of 40 at 0.6% NaCl	addition of saline water 0 to 2.4% NaCl	electrical resistance (Corrosometer)	Griffin and Henry [5]
0.012 to 2.65	addition of 0 to 4%[c] $CaCl_2$ to 1:2.5:3.5 cement	electrical impedence	Dawson et al [6]
0.056 to 0.344	addition of 0 to 5%[c] $CaCl_2$ to slag cement; W/C = 0.18 to 0.7	polarization resistance	Gouda et al [7]

[a]RH = relative humidity.
[b]W/C = water/cement.
[c]Concentration weight percent based on cement content.

concrete). The concrete mix design was suggested by the Oklahoma Department of Transportation. The amount of material which was used for each batch is given in Table 2. Sodium chloride was dissolved in the water and added in the mixing process. The cylinders were then put in a closed cabinet about 2.5 cm (1 in.) above the water level.

In the second set 1.27-cm (0.5 in.) reinforcing steel bars as received were coated with an epoxy resin so that 4.5 cm^2 (6.9 $in.^2$) of their surface area was left bare. The applied coating was the mixture of 2:1 parts of Epon resin 828

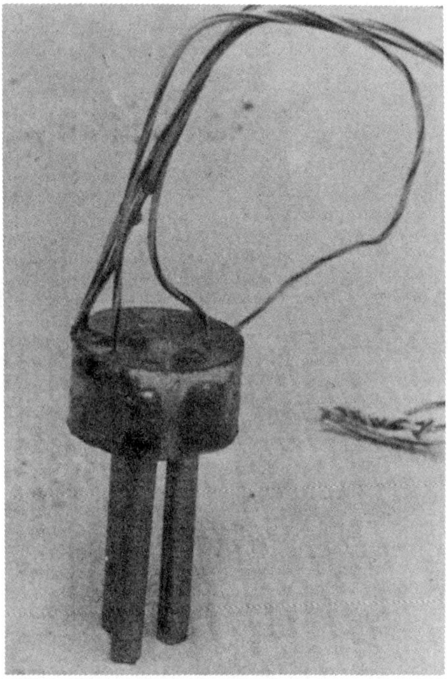

FIG. 1—*Three-electrode probe.*

TABLE 2—*Concrete mix design.*

Cement Type I	10.9 kg (24 lb)
Aggregate	34 kg (75 lb)
Sand	22.7 kg (50 lb)
Water	4.5 kg (10 lb)
Air entrainment	18 ml
Sodium chloride[a]	...

[a]Weight percent of added sodium chloride versus total weight of concrete was 0, 0.05, 0.08, 0.1, 0.2, 0.5, and 1.

and F-5 as the curing agent (both from Shell Co.) with 1.5 weight percent of Cab-O-Sil as thickener. The rebars were cleaned carefully with acetone and the partially coated steel bars were washed with acetone and symmetrically cast, 5 cm (3 in.) apart, in 15.2 by 30.5 cm (6 by 12 in.) concrete cylinders containing 0 to 1 percent sodium chloride. The mill scale existing on the bars then was not removed. The same concrete mix design was used in both sets. All the specimens were stored in the water cabinet for 28 days to cure.

Electrochemical Tests

All polarization tests were carried out using an Aardvark PEC-1B potentiostat. This instrument can be adapted to operate as a galvanostat as was done in some of the experiments. A Keithly Model 602 electrometer and digital voltmeter (HP 3440 A) were used to record potentials. The corrosion rates were compared also with those obtained from a commercial corrosion rate measurement instrument (Petrolite model M-1013).

Machined Probe Specimens

The specimens were taken out of the water cabinet 10 days after they had been made. The experiments were started with polarization resistance tests. The corrosion rates were then recorded by using the commercial-type corrosion rate measurement instrument. Finally, anodic and cathodic polarization tests were conducted. All polarization tests with these specimens were galvanostatic. A $Cu/CuSO_4$ electrode served as the reference electrode. The $Cu/CuSO_4$ was placed against the concrete cylinder with a potassium chloride wetted sponge as a contact point. One of the electrodes of the probe served as the counterelectrode. For the polarization resistance tests the applied current increments at each step were 0.01 to 0.05 μA, lasting for 3 to 5 min until the potentials stabilized. For anodic and cathodic polarization tests the current increments were chosen in such a manner that around $\pm$50-mV shift in potential resulted at each step.

Reinforcing Steel Specimens

The specimens containing partially coated reinforcing steel was cured 28 days in a water cabinet, dried under the laboratory conditions for about a month, and then tested. The polarization resistance tests were carried out galvanostatically on each specimen both anodically and cathodically. These tests were done essentially with the same procedure as for the machined probes. The applied current increments were 0.05 to 2.4 μA, depending on the salt content. The anodic and cathodic polarization tests for these specimens were carried out potentiostatically with $\pm$50-mV increments.

Corrosion Rate Calculation

The results of the polarization resistance tests together with the anodic and cathodic Tafel slopes (β_a and β_c) have been used to calculate corrosion current (i_{corr}) by using the Stern-Geary equation

$$i_{corr} = \left(\frac{di}{dE}\right) \frac{\beta_a \beta_c}{2.3\,(\beta_a + \beta_c)} \tag{1}$$

where di/dE is the slope of the polarization resistance curves at the free corrosion potential ($i = 0$). The corrosion currents were then converted to corrosion rates according to

$$Fe \rightarrow Fe^{++} + 2e \tag{2}$$

and Faraday's law.

This equation was derived assuming that there was only one electrode reaction occurring on the surface and that the electrode had a uniform surface. These conditions may not exist on the surface of the reinforcing bar used with mill scale intact. Therefore, the results of these tests may have some errors due to these factors. However, this should not negate the usefulness of these data in examining the effect of salt content on the corrosion rate in the present experiments.

Results

In order to calculate the corrosion rates, three factors should be determined according to eq 1. These are the slope of the polarization resistance curves, the slope of the anodic polarization curve (β_a), and the slope of the cathodic polarization curve (β_c). In most corrosion rate calculation experiments the values of β_a and β_c are assumed to be in the range of 30 to 120 mV; however, in this investigation all Tafel slopes have been determined experimentally.

Figure 2 shows the results of polarization resistance tests for machined probes which were conducted for a 20-mV shift from the free corrosion potentials. The results of anodic and cathodic polarization tests for these specimens are shown in Figs. 3 and 4, respectively.

All the slopes have been determined graphically and are listed in Table 3 together with calculated corrosion rates. The corrosion rates from the commercial-type instrument have also been included in this table.

Polarization resistance and anodic and cathodic polarization experiments have also been conducted for the specimens with steel rebars. Figure 5 shows the galvanostatic setup for these tests. The results of the polarization resistance tests are shown in Figs. 6 and 7. Two different types of polariza-

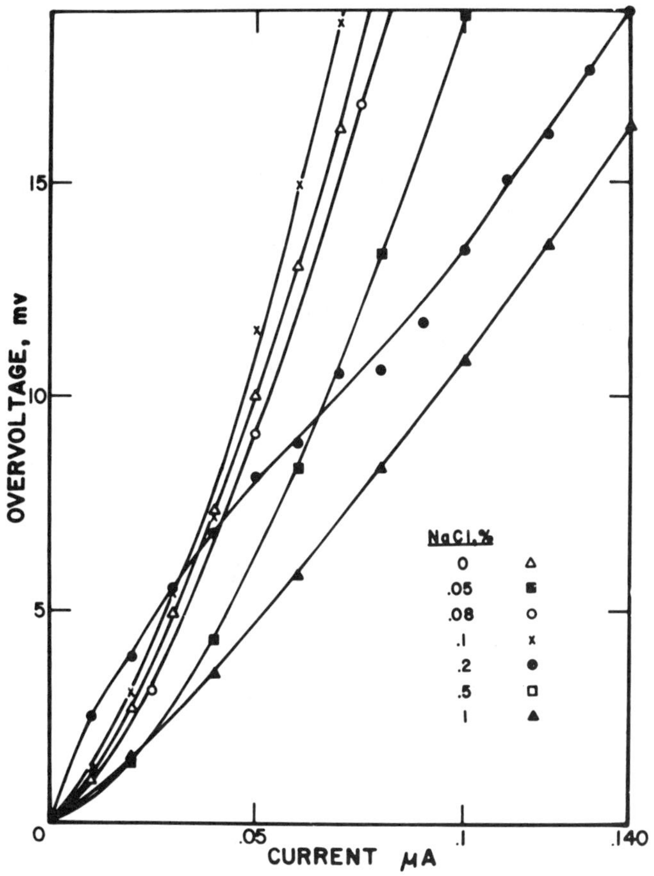

FIG. 2—*Polarization resistance of 3-electrode probe.*

tion resistance experiments were done. In one group of specimens the current was anodic and the potential was increased up to 20 mV above the free corrosion potential. In the other group the current was cathodic and the potential was decreased to 20 mV below the free corrosion potential. In Fig. 8 the anodic and cathodic polarization tests are shown together. The Tafel slopes, the calculated corrosion rates, and the results from the commercial instrument are tabulated in Table 4.

Discussion

As stated earlier, the corrosion process of reinforcing steel in concrete is a complex problem due to the number of variables involved in making the concrete and steel, and the environment as well. The most important factors,

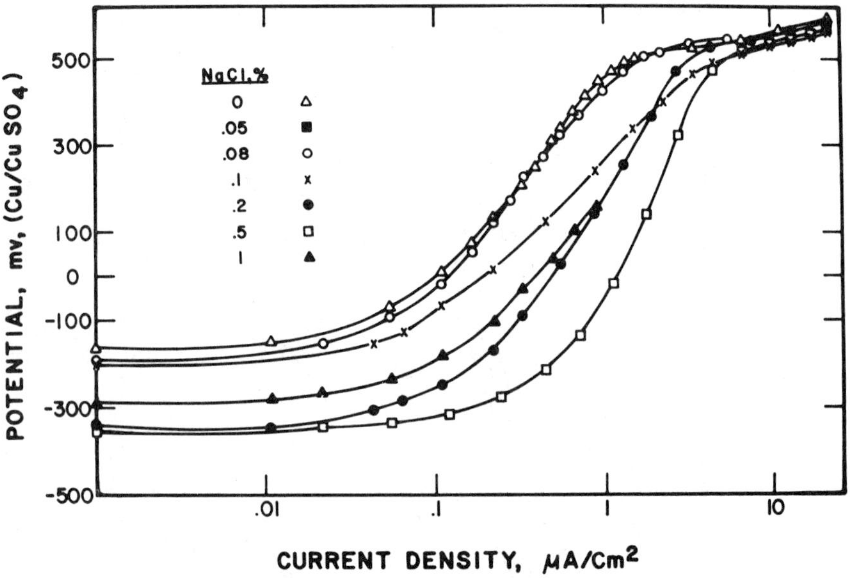

FIG. 3—*Anodic polarization of 3-electrode probe.*

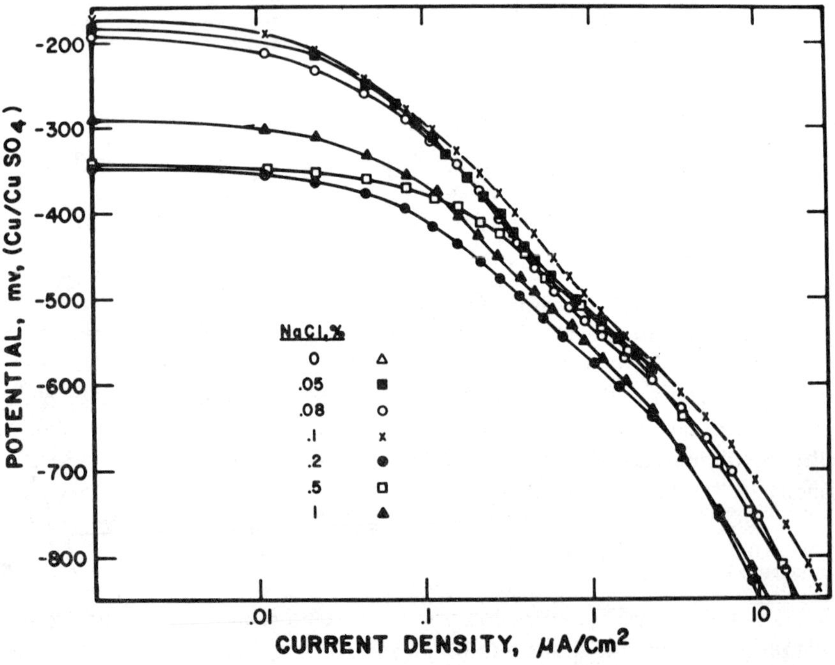

FIG. 4—*Cathodic polarization of 3-electrode probe.*

TABLE 3—*Corrosion rate of machined probes.*

NaCl %	$di/dE \times 10^6$ A/V	β_a, mV	β_c, mV	Corrosion Rate, mils per year	Corrosion Rate: Commercial Instrument, mils per year
0	8.5	530	300	0.036	0.01
0.05	15	500	320	0.064	0.01
0.08	10	490	280	0.039	<.01
0.1	75	400	220	0.023	0.01
0.2	70	600	180	0.021	0.02
0.5	...	830	210	...	0.04 to 0.05
1.0	15	450	190	0.044	0.02 to 0.04

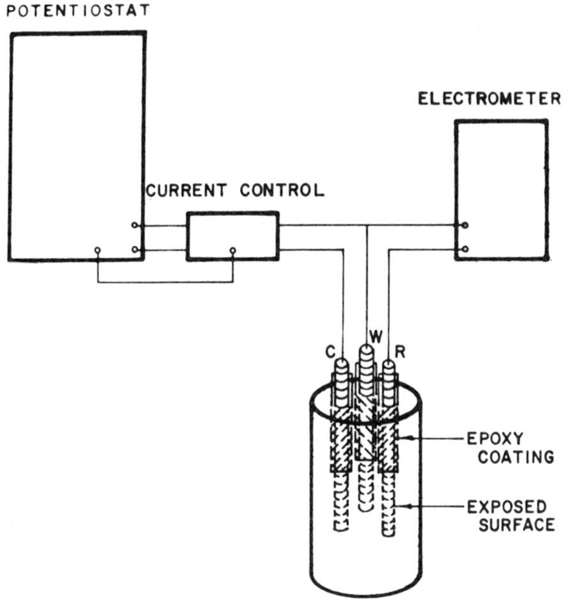

FIG. 5—*Galvanostatic polarization setup for reinforcing steel.*

however, are the surface condition of the steel, moisture content, access to oxygen (permeability of concrete), and the presence of Cl^- ions in the environment and concrete.

The mill scale covering the surface of the steel consists of three layers of iron oxides. Ferrous oxide (FeO) is adjacent to the steel, magnetite (Fe_3O_4) is in the middle, and ferric oxide (Fe_2O_3) on top [*8*]. The electrochemical mechanisms which have been proposed so far are the anodic reaction of iron to Fe^{++} at local anodes and cathodic reduction of oxygen at local cathodes. The existence of mill scale has affected the corrosion mechanism and

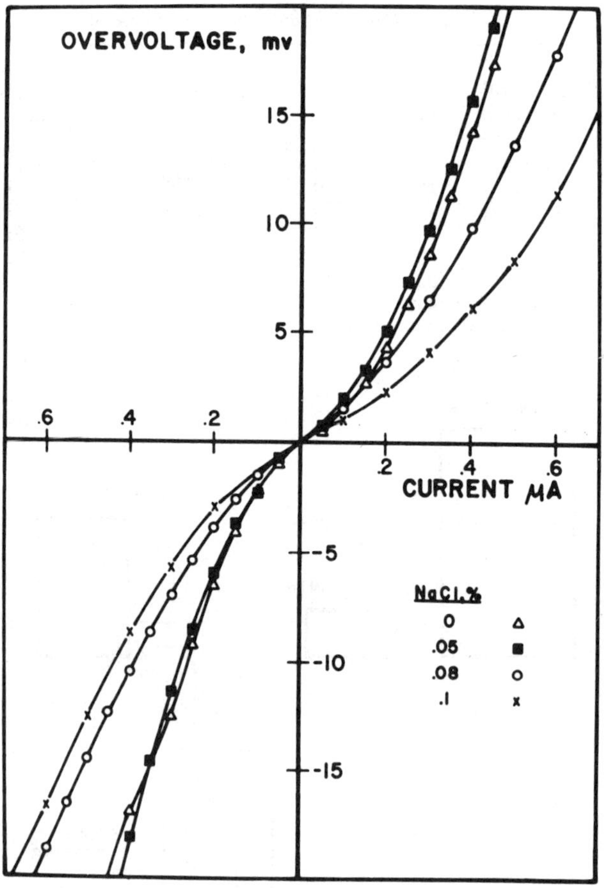

FIG. 6—*Polarization resistance of reinforcing steel for 0, 0.05, 0.08, and 0.1 percent salt.*

polarization curves. In addition to the surface condition effects, moisture content should be taken into consideration. The machined probes were tested 10 days after they were made while the specimens were still wet. At this early stage, as can be seen from Table 3, the corrosion rates are of the same order of magnitude and the chloride has not been effective in changing the surface condition by that time. In an earlier study by the authors, similar results were observed by potential measurements. The moisture content not only affects the cathodic reaction but it also changes the conductivity of concrete.

The importance of Cl^- ions has been better demonstrated in the tests with reinforcing bars. Generally the results of the tests on 0, 0.05, 0.08, and 0.1

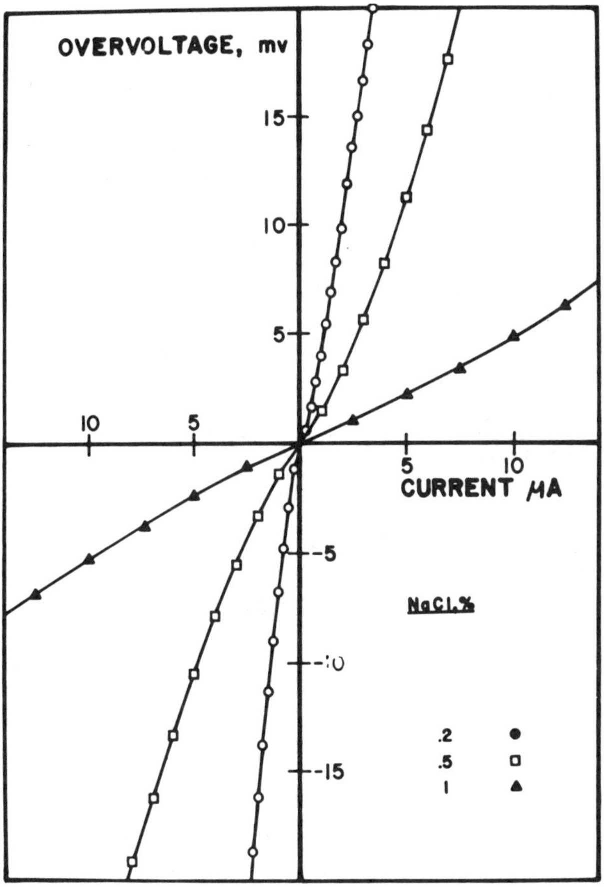

FIG. 7—*Polarization resistance of reinforcing steel for 0.2, 0.5, and 1 percent salt.*

percent salt are almost identical. Reactions reported by Mehta [*9*] and Ben Yair [*10*] of Cl^- with portland cement could remove the chloride and thus maintain the lowered corrosion rate. As can be seen from Table 4, changing the sodium chloride content from 0.1 to 0.2 percent has resulted in a significant increase in corrosion rate. This is in agreement with the chloride concentration threshold theories [*11*]. A value of 0.65 to 0.77 kg/m^3 (1.1 to 1.3 lb/yd^3) of Cl^- ions as the critical concentration in bridge decks has been suggested [*12*]. These results indicate that the threshold value may be higher since 0.1 to 0.2 percent is 1.54 to 3.1 kg/m^3 (2.6 to 5.2 lb/yd^3).

The high values of Tafel slopes calculated in this investigation can be attributed to IR drop due to the high resistivity of concrete. It is generally ac-

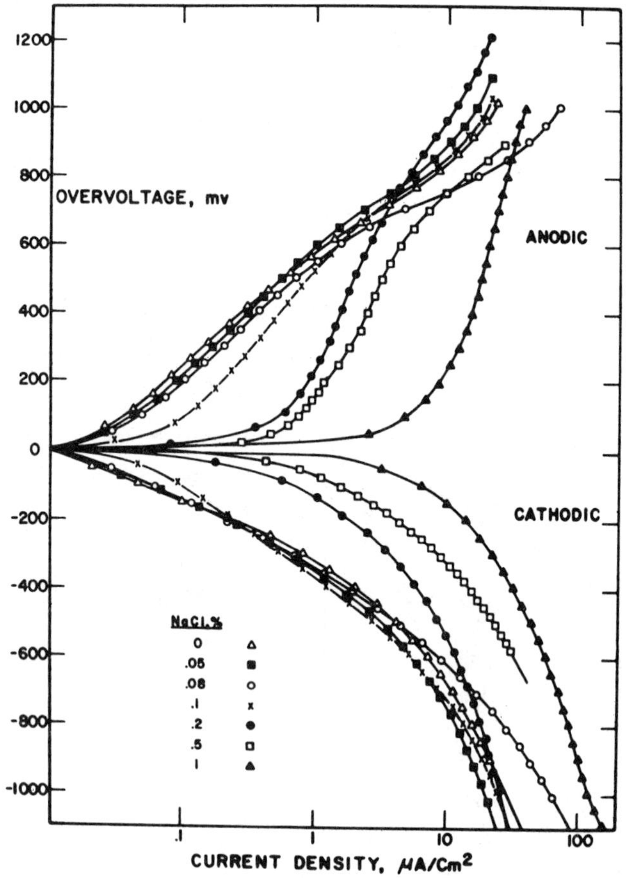

FIG. 8—*Anodic and cathodic polarization of reinforcing steel.*

cepted that Tafel slopes for most materials in various environments are in the range of 30 to 120 mV; however, some exceptions have been reported [*13*]. Jones [*14*] and Lowe [*15*], after working with high-resistivity media and underground buried metals, have proposed a modified circuit for polarization experiments by addition of an electrical bridge to compensate for the IR drop. Mansfeld [*16*] has proposed a positive feedback technique for compensation of IR drops. As long as these effects have not been studied, the results from the commercial-type corrosion rate meters should be accepted with caution. The Tafel slopes ratio $[\beta_a\beta_c/(\beta_a + \beta_c)]$ built in as a constant calibration for the commercial instrument used in the percent research was 81.6 mV, which is far below the experimental values of the investigation. Unfortunately, enough reported data could not be found on Tafel slopes and corro-

TABLE 4—*Corrosion rate of reinforcing steel.*

NaCl %	$di/dE \times 10^6$ A/V	β_a, mV	β_c, mV	Corrosion Rate, mils per year	Corrosion Rate: Commercial Instrument, mils per year
0	100	320	460	0.08	0.003
0.05	80	360	450	0.07	0.003
0.08	70	360	380	0.06	0.005
0.1	100	470	400	0.09	0.006
0.2	250	860	1050	0.52	0.07
0.5	700	850	950	1.39	0.2
1.0	2,000	1570	1080	5.66	0.52

sion rates using the polarization resistance technique in portland cement concrete.

Even though the slopes of the polarization curves may be in error, the comparison of them obtained with different levels of salt content is instructive. The anodic and cathodic polarization curves in Fig. 8 illustrate the chloride threshold concept discussed in the preceding. The anodic polarization curves for sodium chloride contents of 0 to 0.1 percent have a slight indication of passivity and are very similar. The curves of 0.2 to 0.5 percent sodium chloride are somewhat changed with an increase in current and higher slopes. The curve at 1 percent has a much higher value of slope with no deflections or indications of the possibility of passivity.

The cathodic polarization curves indicate little difference in the electrochemical behavior of the 0 to 0.1 percent sodium chloride, consistent with the other results. The current requirements increase with increasing salt content with 0.2 to 1 percent sodium chloride.

Although the existence of IR drops and their effects have not been investigated in this research, the results from the reinforcing bar specimens indicate the importance of Cl^- ions in the corrosion process, their reaction with cement, and the existence of a critical salt concentration. Despite the fact that many variables are involved in studying the corrosion process of steel in concrete, the results of the present investigation are comparable to the results of others reported in Table 1. In addition, other more recent field results have been received. Based on several observations by Hover [*17*] for concrete structures with salt contents of 3 to 6 kg/m^3 (5 to 10 lb/yd^3), the corrosion rates have been estimated to be in the range of 3 to 6 mils per year. In a commercial food processing facility, daily application of $CaCl_2$ for 35 years had resulted in severe deterioration to the structure. The estimated corrosion rate in this case was 6 mils per year. In another case the corrosion rate of several floors of a parking garage exposed to salt was calculated to be around 3 mils per year after 20 years. The results of the present investigation agree fairly well with these field results.

Acknowledgment

This research was supported by the Oklahoma Department of Transportation.

References

[*1*] Cady, P. D. in *Chloride Corrosion of Steel in Concrete, ASTM STP 629,* American Society for Testing and Materials, 1977, pp. 3–11.
[*2*] Raphael, M. and Shalon, R. in *Proceedings,* International RILEM Symposium, Vol. 1, 1971, pp. 177–196.
[*3*] Alekseev, S. N. and Rozental, N. K., *Protection of Metals,* Vol. 10, 1974, pp. 536–538.
[*4*] Akimova, K. M., *Protection of Metals,* Vol. 13, 1977, pp. 157–159.
[*5*] Griffin, D. F. and Henry, R. L. in *Proceedings,* American Society for Testing and Materials, Vol. 63, 1963, pp. 1046–1075.
[*6*] Dawson, J. L., Callow, L. M., Hladky, K., and Richardson, J. A. in *Proceedings,* Corrosion/78, National Association of Corrosion Engineers, 1978, Paper 125.
[*7*] Gouda, V. K., Shater, M. A., and Mikhail, R. S., *Cement and Concrete Research,* Vol. 5, No. 1, 1975, pp. 1–13.
[*8*] Shrier, L. L., *Corrosion,* 2nd ed., Newnes-Burtterworths, London, U.K., 1976.
[*9*] Mehta, P. K. in *Chloride Corrosion of Steel in Concrete, ASTM STP 629,* American Society for Testing and Materials, 1977, pp. 12–19.
[*10*] Ben-Yair, M., *Cement and Concrete Research,* Vol. 4, No. 3, 1974, pp. 405–416.
[*11*] Hausmann, D. A., *Materials Protection,* Vol. 6, No. 11, 1967, pp. 19–23.
[*12*] Clear, K. C. and Hay, R. E., Federal Highway Administration Report No. FHWA-RD-73-32, Vol. 1, 1973.
[*13*] Becerra, A., and Darby, R., *Corrosion,* Vol. 30, No. 5, 1974, pp. 153–160.
[*14*] Jones, D. A., *Corrosion Science,* Vol. 8, No. 1, 1968, pp. 19–27.
[*15*] Jones, D. A. and Lowe, T. A., *Journal of Materials,* Vol. 4, No. 3, 1969, pp. 600–617.
[*16*] Mansfeld, F., *Advances in Corrosion Science and Technology,* Vol. 6, Plenum Press, New York, 1976.
[*17*] Hover, K. C., THP Consulting Engineers, Private communication, Jan. 1979.

I. Cornet,[1] D. Pirtz,[1] M. Polivka,[1] Y. Gau,[2] and A. Shimizu[3]

Laboratory Testing and Monitoring of Stray Current Corrosion of Prestressed Concrete in Seawater

REFERENCE: Cornet, I., Pirtz, D., Polivka, M., Gau, Y., and Shimizu, A., **"Laboratory Testing and Monitoring of Stray Current Corrosion of Prestressed Concrete in Seawater,"** *Corrosion of Reinforcing Steel in Concrete, ASTM STP 713*, D. E. Tonini and J. M. Gaidis, Eds., American Society for Testing and Materials, 1980, pp. 17–31.

ABSTRACT: Stray current corrosion of prestressed concrete beams was investigated in the laboratory by exposing 40 specimens 6.4 by 6.4 by 122 cm (2.5 by 2.5 by 48 in.), prestressed by a central high-strength steel wire to 1.86×10^9 N/m^2 (270 ksi), in seawater. The steel wire was made anodic to a copper cathode, with steel current densities maintained at fixed values between 27 and 915 mA/m^2 (2.5 and 85 mA/ft^2).

Monitoring was done by measuring steel potential relative to a silver/silver chloride reference electrode with current on, weekly, and with current off, biweekly. Beams were examined visually biweekly; the presence of rust spots and longitudinal cracks was noted, and lengths of cracks were measured, for exposures which ranged between 8 and 83 weeks. After exposure, the prestressing wire was tensioned to failure.

Reductions in breaking strength of 70 percent were observed in 25 weeks' exposure at 915 mA/m^2 (85 mA/ft^2), with lesser reductions in strength for shorter exposures and lower current densities.

Ampere-hours did not correlate satisfactorily with the reduction in breaking strength of the wire. Potentials measured with current on or off indicated that corrosion was occurring, but gave no quantitative indication of the reduction in breaking strength. Resistance measurements of the electrochemical circuit did not relate to the extent of corrosion damage. Time to change in potential of the prestressing steel did correlate with time for initiation of steel corrosion. Existence and length of longitudinal cracks in the concrete beam did not correlate quantitatively with the reduction in breaking strength of the prestressing steel.

After the tension test, beams were notched lengthwise with a saw and opened. The prestressing wire was then examined to determine the distribution and extent of corrosion. Quantitative estimates of the corroded length were made. Qualitatively, where there was considerable localized corrosion attack, there was great reduction in breaking strength for a given number of ampere-hours' exposure. Where the

[1] Professor emeritus, Department of Mechanical Engineering, and professors, Department of Civil Engineering, University of California, Berkeley, Calif. 94720.

[2] Engineer, Union Carbide Corp., Bound Brook, N.J. 08805.

[3] Assistant manager, Coen Co., Burlingame, Calif., 94010.

corrosion attack was well distributed, an equal number of ampere-hours gave less reduction in fracture strength.

Stray electrical currents can cause serious deterioration in the strength of prestressed concrete structures, as measured by testing to destruction. However, none of the methods of monitoring used in this investigation can predict the extent of the damage quantitatively.

KEY WORDS: laboratory testing, monitoring, stray current corrosion, prestressed concrete in seawater

Steel under tension in concrete in seawater is susceptible to corrosion by stray anodic electric currents.

The passivity of steel in an alkaline environment is well known. Concrete provides such an environment for embedded steel. Breakdown of passivity can occur due to a lowering of the pH under the action of stray electrical current [*1*][4] in seawater.

Steel starts to corrode when the pH is less than 11.5 with oxygen and water present [*2*].

The critical chloride concentration for steel to start to corrode depends on various factors. As low as 700 ppm of chloride in concrete [*3*] in the presence of oxygen can start corrosion. In the absence of oxygen, the threshold chloride content is about four times higher at a potential of about −0.4 to −0.5 V relative to a saturated calomel electrode [*4*]. This would suggest that the specification of the pH and the chloride concentration to corrosion should be supplemented by the specification of the steel potential, because a relation may exist among the three quantities involved. The critical level of chloride was also found to be a function of the cement factor and the water/cement (W/C) ratio [*5*].

The presence of chloride itself does not affect markedly the pH of the concrete [*6*].

The mechanism of steel corrosion in concrete is reasonably well understood. Oxygen is needed at the cathodic area for the reaction $2H_2O + O_2 + 4e^- \rightarrow 4\,OH^-$ to go, and a minimum content of the aggressive ions, that is, chloride ions, is required to break down the passive film with the resulting dissolution of iron $Fe \rightarrow Fe^{++} + 2e^-$, at the anode.

Diffusivities of dissolved oxygen through concrete pores have been measured [*7*], but these values will be affected by the degree of oxygen saturation of concrete.

Chloride ions will move into concrete by two transport mechanisms, diffusion and convection or moisture motion. The reaction between chloride ion and tetracalcium aluminate in concrete [*8*] renders the analysis of chloride diffusion more complicated.

The foregoing corrosion process is no longer valid for steel in concrete

[4]The italic numbers in brackets refer to the list of references appended to this paper.

immersed in seawater and subjected to an external d-c current source. In the early stage where seawater has not penetrated into the concrete, the free calcium hydroxide in solution in the free moisture of concrete will provide most of the elements for the reactions at the anode and cathode

$$O_2 + 2H_2O + 4e^- \underset{\text{anode}}{\overset{\text{cathode}}{\rightleftharpoons}} 4OH^-$$

Hydrogen evolution occurs at the cathode only at a potential less than −1.1 V silver/silver chloride (Ag/AgCl) in seawater at 25°C (77°F).[5] At the same time, anions, mostly Cl^-, and cations, Na^+ and Ca^{++}, in concrete are attracted toward the anode and cathode, respectively. In an electrolysis test [*9*] of concrete in sodium chloride solution, traces of calcium ion were found in the solution.

In the absence of corrosive environments, the anode steel corrodes when the hydroxyl ions are sufficiently depleted at the steel concrete interface. There exists then an induction time. Chloride ion will shorten the induction time. It was found that 300 ppm of chloride [*10*] is sufficient to promote corrosion. When concrete is 100 percent saturated with seawater, the mechanism of ion motion is more complex and needs further study.

Once corrosion starts, the dissolution of iron does not take place all along and around the steel surface and may not be the only reaction at the anode. In part this is a function of anode efficiency, as well as localized corrosion of the steel. The distribution of the corrosion will have a crucial importance in determining the strength left in the steel.

Methods of monitoring corrosion of steel in prestressed concrete are visual inspection, resistance, and potential measurement. The present laboratory work was undertaken for assessing the method of monitoring and testing of stray current corrosion of prestressed concrete in seawater, seeing how well it could detect the corrosion in its different stages, and evaluating the corrosion damage to the embedded steel.

Test Specimens

Beams 63.5 by 63.5 by 1219 mm (2.5 by 2.5 by 48 in.) with a single prestressing wire centered within a square cross section were used. The prestressing wire was the center wire of an uncoated seven-wire stress-

[5]The Ag/AgCl reference electrode was made of a piece of 99.97 percent pure silver wire made anodic, cathodic, and then anodic in 0.1-N hydrochloric acid (HCl). After rinsing and exposure to seawater, such electrodes are reproducible and are within 8 mV of a saturated calomel electrode at 25°C (77°F). Theoretically the potential at which H_2 is released in a medium of pH 12.4 is −0.0592 pH = −0.734 V standard hydrogen electrode or −0.98 V Ag/AgCl. However, as the cathode becomes more alkaline and as there are overvoltage effects, a potential of −1.1 V Ag/AgCl is adopted.

relieved strand for prestressed concrete conforming to ASTM 416-68 Grade 270 specification. The average strength was 1.86×10^9 N/m^2 (270 ksi) and the average diameter 4 mm (0.172 in.).

Two wires were prestressed to 1.2×10^9 N/m^2 (175 000 $\pm$ 5000 psi) between two reinforcing steel floor anchors prior to the day of casting. The load was applied with a hydraulic jack at one end and monitored at the other end by a transducer load cell connected to a strain-gage indicator. Six beams were made at a time in three wooden forms. After casting, the beams were cured with moist burlap for seven days. Then the prestress force was transferred to the concrete. Beams were cured seven more days in a fog room and 14 additional days in dry air. During the last week of air curing, 22.8 cm (9 in.) of the beam ends were coated twice with epoxy resin "Concrete Concresive 1170" to suppress end effects. This epoxy resin was also applied to the protruding ends of wire, which were further protected by encasing in vinyl tubing and sealing with a marine-type silicon sealant. Copper wire of 3 mm (0.125 in.) diameter served as a cathode installed in either diffuse (Z-wrapped) or concentrated (single loop) configuration. The two geometries for the cathode were for simulating the effect of the current distribution (Fig. 1). Concrete mix proportions and compressive strength are given in Table 1.

Monitoring and Testing

Each beam was put into an individual tank of 152-mm (6 in.) inside diameter and 1524-mm (60 in.) height filled with synthetic seawater of composition given in Table 2, in a room kept at 15 $\pm$ 1°C (60 $\pm$ 2°F). The prestressing steel was connected to the plus terminal and the copper

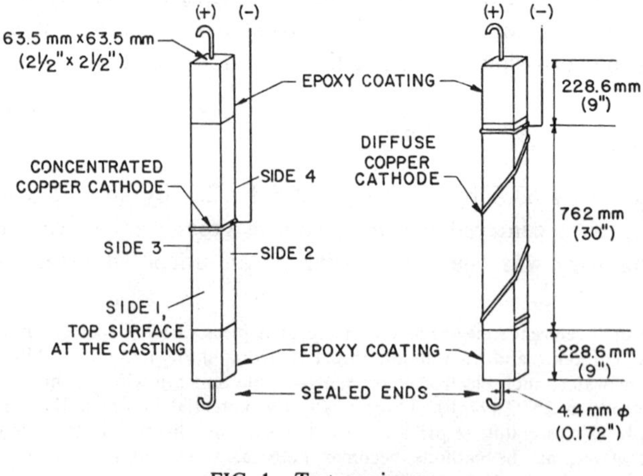

FIG. 1—*Test specimens.*

TABLE 1—*Concrete data.*

Component	Proportions	
	kg/m³	(lb/yd³)
Cement Santa Cruz Type II	390.37	(658)
Water	175.01	(295)
Coarse aggregate Fair Oaks, 6.35 to 12.7 mm (1/4 to 1/2 in.)	886.35	(1494)
Fine aggregate	956.36	(1612)
Compressive Strength	N/m²	(lb/in.²)
7 days	2.654×10^7	(3850)
9 days	2.930×10^7	(4250)
28 days	3.516 to 4.137×10^7	(5100 to 6000)

TABLE 2—*Composition of synthetic seawater.*

Component	Grams per 100 litres of Solution
Calcium chloride	220
Magnesium chloride	1100
Potassium bromide	90
Potassium chloride	20
Sodium chloride	2300
Sodium sulfate	353

wire to the minus terminal of the power supply. The maximum potential used was 3 V. The rectified current was maintained at its set value daily during weekdays by means of a resistor in series with the beam. A schematic of the electric circuit is shown in Fig. 2. Potential measurements with current on were performed every week with the steel anode connected to the plus terminal of a millivoltmeter (100-MΩ input impedance) and the Ag/AgCl reference electrode to the minus terminal; resistance and potential measurements were made with current off, biweekly. The measured potential obtained in this way corresponds to the reduction potential in the thermodynamic sign convention. For the resistance measurement, an a-c meter (Vibroground Model 293, two-points method) was used. In the early part of the experiment, the resistance was measured between the copper cathode and the embedded steel. Due to the calcareous coating depositing on the cathode, a foot of steel wire similar to the anode replaces the copper cathode in later measurements. For some beams, the obtained values were compared with those measured with a regular ohmeter. Results are based on the a-c meter measurement.

After a specified time of exposure, beams that had shown visible signs

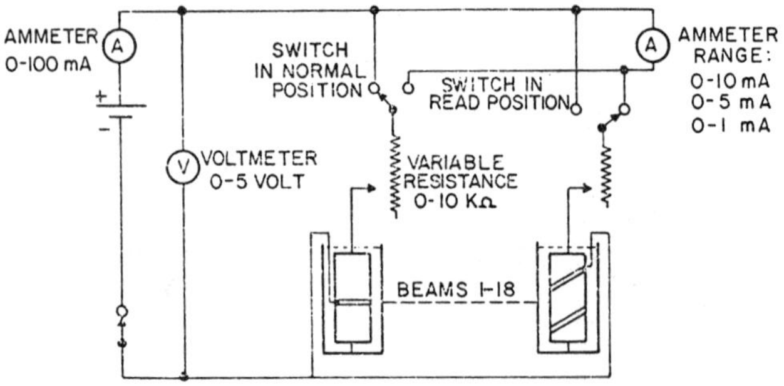

FIG. 2—*Electric circuit.*

of corrosion were subjected to testing in tension to failure. The test required almost the same equipment as for prestressing wire. A transducer load cell was connected to the *X*-axis of an Esterline Angus *XY* plotter and a time scale generator to the *Y*-axis. Data were obtained when the beam cracked transversely and when the wire broke. After the test, the beams were sectioned longitudinally, and the wire examined.

Results

Visual inspection has been widely used for detecting steel corrosion in concrete. It is not too practical for fully submerged structures. Corrosion products larger in volume than the volume of steel replaced cause cracks which may precede or follow rust staining. If the buildup of internal pressure exceeds the breaking strength before rust reaches the concrete surface, cracks precede rust stains. Sometimes rust stains appeared before cracks; sometimes cracks and rust could be seen at the same time. An example of the crack and rust propagation is shown in Fig. 3. The presence of a crack or rust spot indicates that steel has been corroding, but will not indicate for how long the steel has been corroding and the extent of corrosion damage. A small rust spot or crack may be associated with an early stage of corrosion, but a small crack with many ampere-hours will lead to a localized attack and a great reduction in breaking strength. When the steel is removed from the concrete, in the early stage of corrosion, only a sector of the steel surface, varying from one quarter to one half on the circumference, is corroded. As the ampere-hours increase, the rusted sector covers more and more surface and finally encircles the whole circumference. The same tendency is true for the corroded length, short for small ampere-hours and more distributed for large ampere-hours. Due to the wide variation in size and depth of pits, no quantitative criterion has been established. The steel under the portion of the insulated concrete remains bright.

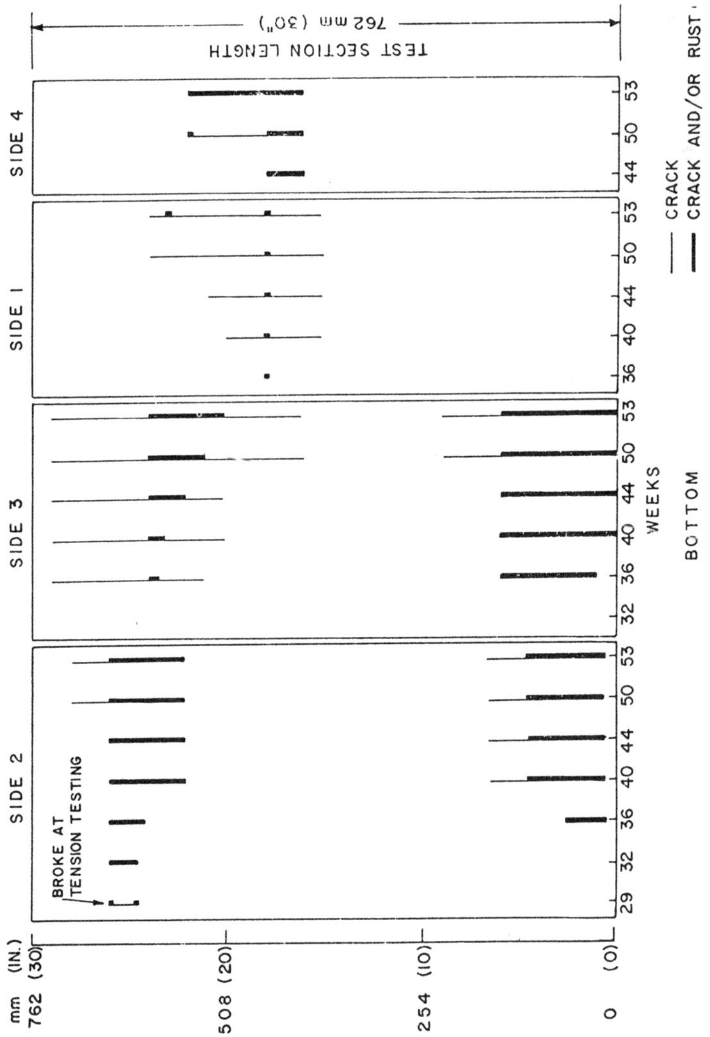

FIG. 3—*An example of crack and rust propagation with time of exposure: beam V 13, diffuse cathode, 215 mA/m² (20 mA/ft²).*

This suggests covering the concrete structure by coatings, but this may not be too practical, and any defects in the cover may shorten the life of the structure prematurely due to a localized attack.

Table 3 and Fig. 4 give an example of the variation of the resistance values with exposure time. The decrease of resistance in the early minutes of exposure is associated with the larger surface area of the concrete in contact with seawater and lasts from one to two weeks for beams with applied current.

For beams without applied current, the decrease in resistance persists up to four weeks before a rise in resistance is noted. After the first or second week's drop in resistance, the beams subjected to stray current showed a steady increase in resistance until the first sign of corrosion, followed by a slow decrease afterwards. It is appropriate at this stage to look at the various factors affecting resistance in porous media. Experiments [*11*] have shown that a homogeneous mixture of conductive solid and electrolyte can be treated as a mixture of two electrolytes with their resistivities weighted, respectively, with their volume fractions.

For a more complete analysis, one has to include the ionic conduction through the saturating bulk fluid, and through the surface or electrical double layer [*12*]. It can be readily seen that the assumption of the current conduction by parallel paths through the conducting solid and the interstitial electrolyte is not at all adequate. The conducting media are not continuous everywhere and interconnected. The resistivity of porous rock is then only a fraction of the resistivity of the mixture. From a purely empirical point of view, the resistivity of porous media may be expressed as [*13*]

$$r_{\text{por}} = \left(\frac{1}{S_w}\right)^2 r_o P^{-m}$$

where

S_w = water saturation,
r_o = resistivity of the rock 100 percent saturated with an electrolyte,
P = porosity of the rock, and
m = a constant depending on the rock.

A similar relation may hold for concrete. If the measurement is taken between two electrodes with the concrete immersed in a neutral electrolyte of low resistivity

$$R_{\text{concrete}} = \left(\frac{1}{S_w}\right)^2 r_o P^{-m} \frac{l}{A}$$

where l is the actual length and A the actual cross-sectional area of the current path. With other variables constant, R_{concrete} decreases as the water

TABLE 3—*Resistance change after exposure to seawater.*

Time of immersion, min	1	5	15	270
Resistance, Ω	76	71	70	62

NOTE: Beam V 12, aged 52 days.

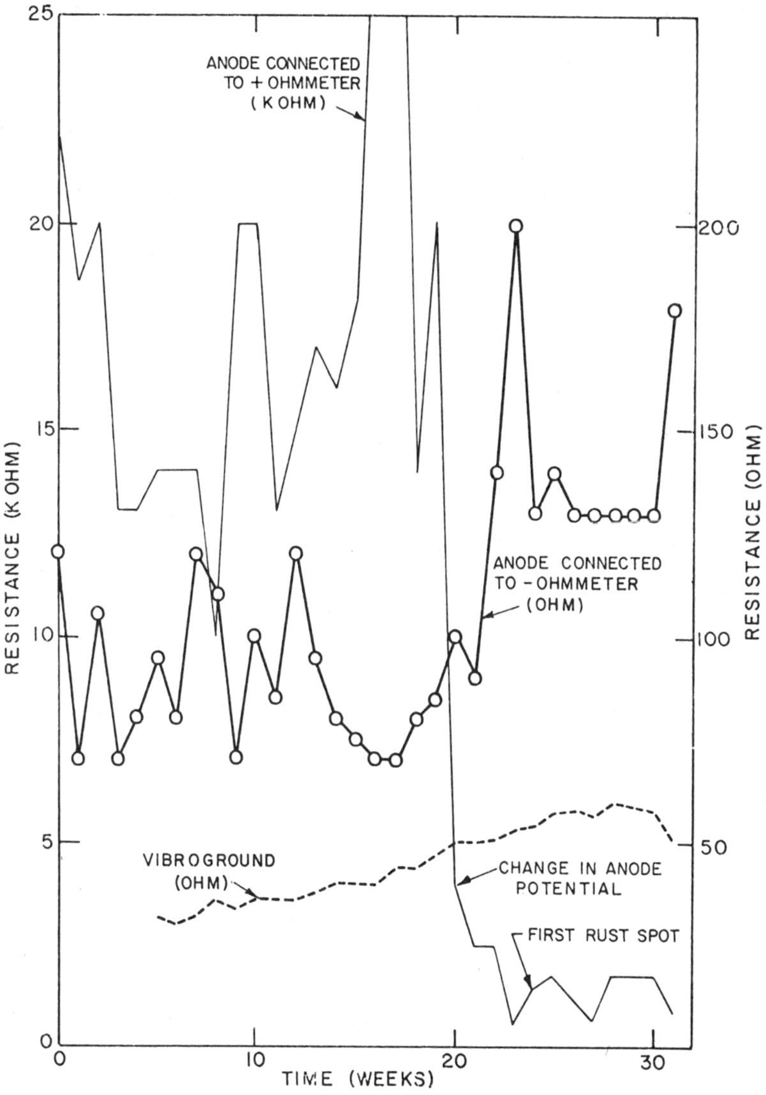

FIG. 4—*An example of variation of resistance with time of exposure: beam III 13, concentrated cathode, 54 mA/m² (5 mA/ft²).*

saturation increases. When concrete is cast against one of the electrodes, the resistance measurement will include the effect of the oxidation or reduction of ions at the steel/concrete interface [*14*]. Below a critical voltage, the resistance appears to be high and varies with the applied voltage. Above it, the ratio voltage/current is essentially constant. The increase of the resistance of concrete with exposure time has been explained as a result of the formation of film due to the passage of current at the steel/concrete interface [*9*] or as a result of the hydration process which reduces the pore systems in the concrete. Pore systems may give room for the transport of ions [*15*]. From the foregoing relation, it is seen that the $R_{concrete}$ is a function of S_w, P, m, l, and A. The degree of saturation S_w is a factor which may have a large influence in the early stage of exposure. With longer time exposure, the reduction in the pore system will have a stronger effect, assuming that m, l, and A do not vary markedly. The resistance of concrete then increases. The decrease in resistance after the beams have reached the time to visible sign of corrosion is associated with a shorter path of the electric current through the seawater electrolyte to the steel surface.

Unless the resistance of the immersed concrete structure is taken periodically from the first day of immersion, the chance of detecting the corrosion activity of steel in concrete is low, and even with a continuous monitoring the method may not be reliable. Potential measurements are a lot safer. Several investigators [*3,6,16,17*] have shown that the half-cell potential of steel with respect to a reference electrode is a good indication of the corrosion activity of steel in concrete. Measurement of half-cell potential [*16*] has identified steel as noncorroding when a measured value is more positive than −0.23 V Ag/AgCl in seawater, and corroding when a value is numerically greater than −0.28 V Ag/AgCl in seawater, for embedded steel in concrete not subjected to stray electric current.

The steel potential with current on (potential on) in this experiment is at first positive to the Ag/AgCl in seawater (Fig. 5). A continuous decrease of the potential on from positive to negative value is an indication that steel is corroding, and the time of depassivation of the steel is picked at the first week of the drop in value of the potential on and called "time to corrosion." This is confirmed by the potential with current off, subsequently called "potential off," and visual inspection. With a few beams, especially those at high current density, this time coincides with the time to visible corrosion (crack or rust spot or both) (Fig. 6). Some beams show a sharp drop of the potential with current on, while for others the decrease in potential follows a smoother path. One of the explanations of the significant drop in potential on only after corrosion had started is that probably oxidation of hydroxyl ions and iron dissolution occur simultaneously for some time before the latter predominates.

The potential off at the time to corrosion varies from beam to beam, with current density, but drops to a value less than −0.3 V versus Ag/AgCl in seawater about a week after the time to corrosion. There exists a large

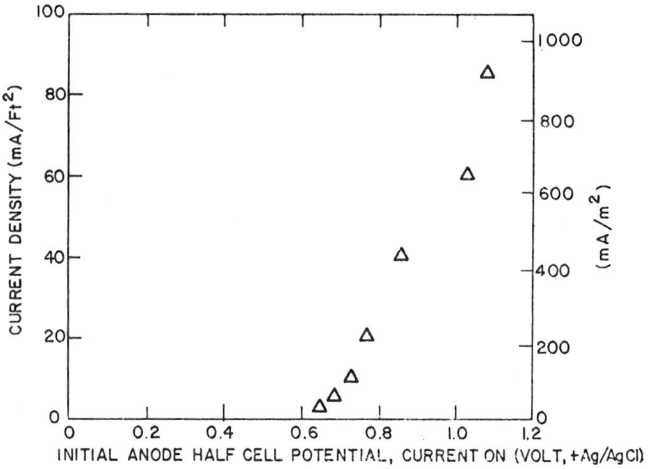

FIG. 5—*Initial half-cell potential as a function of current density.*

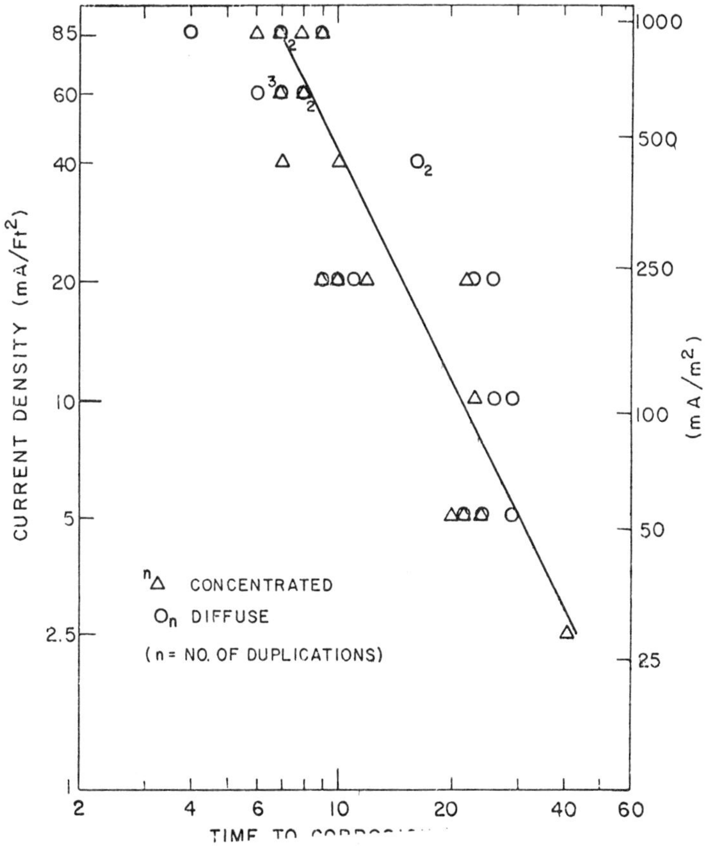

variation of the leveling-off value of the potential on. The leveling of the potential with current off is around −0.5 V versus Ag/AgCl in seawater. The maximum reported value of half-cell potential off is −0.6 V. The highest value obtained in this experiment is −0.54 V. When steel in concrete is first submerged in seawater, the steel potential is around −0.1 V; the potential of bare steel in seawater is −0.43 V versus Ag/AgCl.

Beams that have passed the time to corrosion will show visible signs of corrosion a few weeks later; the lower the current, the longer the time between the change in potential on and the appearance of a crack or rust spot or both (Fig. 7). For both Figs. 6 and 7, a straight-line fit is drawn between the current densities 915 and 27 mA/m^2 (85 and 2.5 mA/ft^2). Beyond these two limits, a straight-line fit may not be correct. The latest data for two beams without impressed current show that a change in steel potential has occurred after 53 weeks of exposure in seawater.

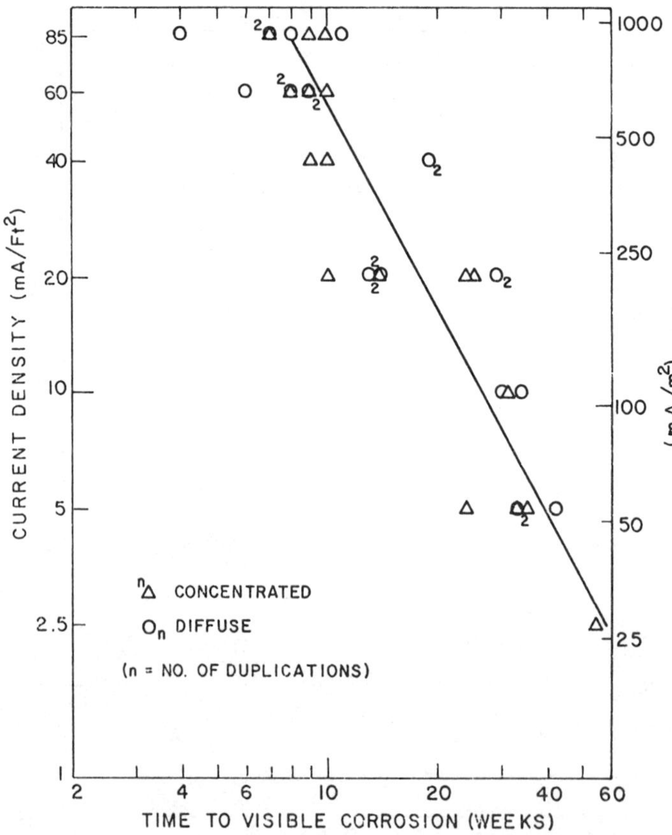

FIG. 7—*Influence of current density on time to visible corrosion.*

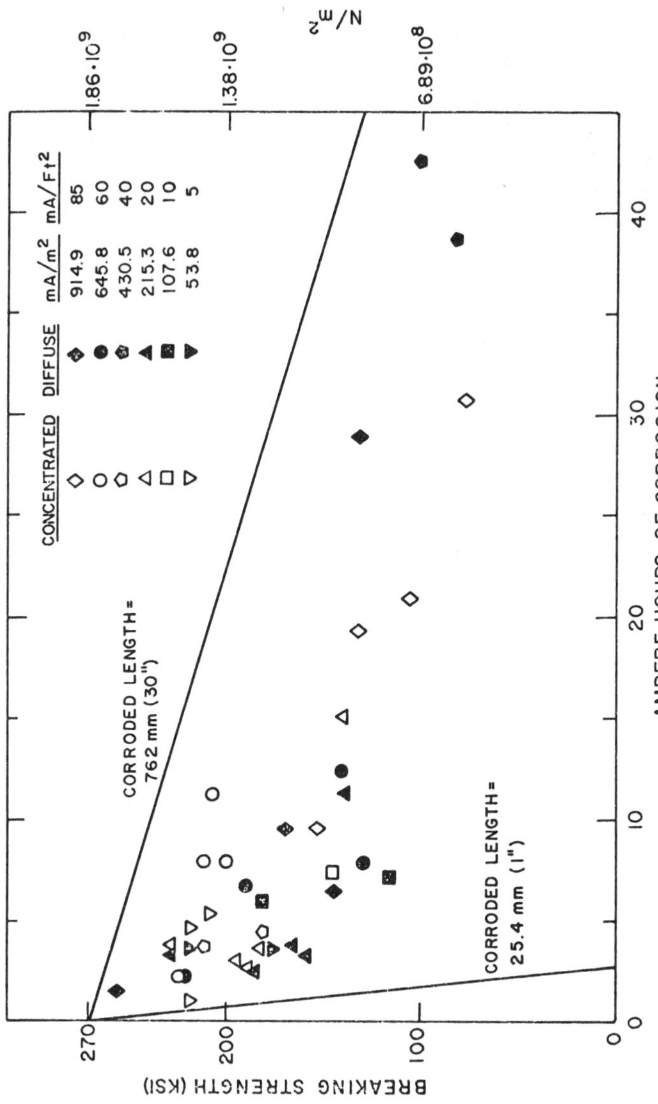

FIG. 8—*Breaking strength versus ampere-hours to corrosion.*

As the potential measurement does not indicate the extent of corrosion damage, beams with visible signs of corrosion were subjected to tension testing to failure after a specified time of exposure. The reduction in breaking strength is shown against the ampere-hours of corrosion in Fig. 8. Each point corresponds to a theoretical corroded length, assuming that corrosion is limited to only a portion of the wire and that the only reaction which has taken place is the iron dissolution. An attempt to compare calculated and real corroded length has been made. The real corroded length is obtained from visual evaluation by estimating the length of pits and the general attack portion. Quantitative conclusions cannot be drawn due to the uneven distribution of the corrosion along the wire. Two lines are shown on Fig. 8, the upper one representing the situation in which corrosion is distributed over 762 mm (30 in.) of the wire (total length of the test section), the lower one representing the reduction in breaking strength with only 25.4 mm (1 in.) of the wire corroded.

Figure 8 may be replotted as fraction of the original breaking strength versus time of corrosion divided by total time. This plot is suggested by the fact that the longer the time of exposure, the longer the corroded length, and the less is the reduction in breaking strength for the same ampere-hours of corrosion (Fig. 9). Two curves are drawn for cases where beams are subjected to a current density of 54 mA/m^2 (5 mA/ft^2) with a corroded length of 25.4 mm (1 in.) and 762 mm (30 in.).

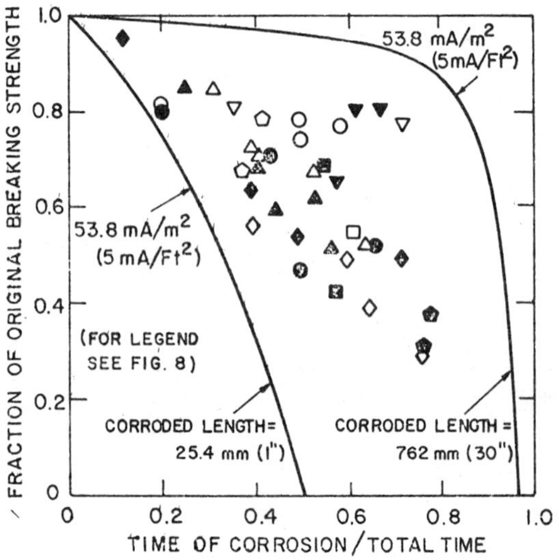

FIG. 9—*Fraction of original breaking strength versus time of corrosion/total time.*

Conclusions

Stray electric current can cause serious deterioration of the strength of prestressed concrete structures.

Reductions in breaking strength of 70 percent were observed in 25 weeks of exposure at 915 mA/m^2 (85 mA/ft^2).

Potential measurement with current on can be used to detect the time to corrosion.

When the potential with current off is less than −0.3 V Ag/AgCl in seawater, steel is corroding.

Circuit resistance measurements are not suitable indicators of corrosion in the steel reinforcement.

None of the monitoring methods used in this laboratory investigation tell the extent of corrosion damage. Only tension testing to failure measured the damage quantitatively.

References

[*1*] Cornet, I., et al in *Proceedings*, Offshore Technology Conference, Houston, Tex., OTC 3195, May 1978.

[*2*] Shalon, R. and Raphael, M. in *Proceedings*, American Concrete Institute, Vol. 55, 1959, pp. 1251–1258.

[*3*] Hausmann, D. R., "Studies of the Mechanism of Steel Corrosion in Concrete," National Association of Corrosion Engineers Western Regional Conference, Honolulu, Hawaii, No. 10, 1965.

[*4*] Ishikawa, T., Cornet, I., and Bresler, B. in *Proceedings*, Fourth International Congress on Metallic Corrosion, Amsterdam, The Netherlands, Sept. 1969, pp. 556–559.

[*5*] Clear, K. C., "Time to Corrosion of Reinforcing Steel in Concrete Slabs," Federal Highway Administration, FHWA-RD-76-70, Vol. 3, April 1976.

[*6*] Berman, H. A. and Chaiken, B., *Public Roads*, March 1976, pp. 158–162.

[*7*] Gjorv, O. E. in *Proceedings*, The International Corrosion Forum, National Association of Corrosion Engineers, Houston, Tex., March 1976.

[*8*] Heller, L. and Benyair, H., *Journal Applied Chemistry*, Vol. 16, Aug. 1966, pp. 223–226.

[*9*] Unz, M., *Corrosion*, Vol. 16, No. 7, July 1960, pp. 115–125.

[*10*] Lewis, D. A. in *Proceedings*, First International Congress on Metallic Corrosion, London, U.K., April 1961, pp. 547–555.

[*11*] DeWitte, L., *The Oil and Gas Journal*, Aug. 1950, pp. 120–132.

[*12*] Olaf Pfannkuch, H. in *Proceedings*, First International Symposium on Fundamentals of Transport Phenomena in Porous Media, Haifa, Israel, 1969; Elsevier, New York, 1972, pp. 42–54.

[*13*] Finley, H., *Corrosion*, Vol. 17, March 1961, pp. 104t–108t.

[*14*] Hausmann, D., *Journal of the American Concrete Institute*, Feb. 1964, pp. 171–188.

[*15*] Bernhardt, C. and Sopler, B., *Nordisk Betong*, Vol. 2, 1974, pp. 22–32.

[*16*] Stratfull, R. F., *Highways Research Record*, No. 433, 1973.

J. C. Griess[1] *and D. J. Naus*[1]

Corrosion of Steel Tendons Used in Prestressed Concrete Pressure Vessels

REFERENCE: Griess, J. C. and Naus, D. J., "**Corrosion of Steel Tendons Used in Prestressed Concrete Pressure Vessels,**" *Corrosion of Reinforcing Steel in Concrete, ASTM STP 713,* D. E. Tonini and J. M. Gaidis, Eds., American Society for Testing and Materials, 1980, pp. 32–50.

ABSTRACT: The purpose of this investigation was to determine the corrosion behavior of a high-strength steel [Specifications for Uncoated Seven-Wire-Stress-Relieved Strand for Prestressed Concrete (ASTM A 416–74, Grade 270)], typical of those used as tensioning tendons in prestressed concrete pressure vessels, in several corrosive environments, and to determine the protection obtained by coating the steel with two commercial petroleum-base greases or with portland cement grout. In addition, the few reported incidents of prestressing steel failures in concrete pressure vessels used for containment of nuclear reactors were reviewed. The susceptibility of the steel to stress corrosion cracking and hydrogen embrittlement and its general corrosion rate were determined in several salt solutions. Wires coated with the greases and grout were soaked for long periods in the same solutions and changes in their mechanical properties were subsequently determined. All three coatings appeared to give essentially complete protection; however, flaws in the grease coatings could be detrimental, and flaws or cracks less than 1-mm-wide (0.04 in.) in the grout were without effect.

KEY WORDS: prestressing steel, high-strength steel, grout, petroleum-base greases, stress-corrosion cracking, hydrogen embrittlement, corrosion, protective coatings

Prestressed concrete pressure vessels (PCPV's) for nuclear reactor containment are massive structures. They are constructed of relatively high-strength concrete which is heavily reinforced by both conventional steel and a steel posttensioning system consisting of vertical tendons and circumferential wire-strand windings. Performance requirements for PCPV's require that extremely large-capacity prestressing tendons fabricated from high-strength steels be utilized to reduce the concentration of steel as much as possible. The wires or strands used to make up the prestressing systems are often small

[1]Engineer, Metals and Ceramics Division and engineer, Engineering Technology Division, respectively, Oak Ridge National Laboratory, Oak Ridge, Tenn. 37830.

in diameter [6 to 7 mm (0.24 to 0.28 in.)] and are used at stresses up to 75 percent of their ultimate tensile strength (UTS). These two facts make corrosion protection of the tendons imperative if the PCPV's are to be structurally safe throughout their 30- to 40-year design life.

Present practice for protecting the tendons in PCPV's is to fill the conduits which contain the tensioned tendons with either a portland cement grout or an organic substance composed of petroleum-based greases or waxes which contain specific additives. Although steels are normally quite compatible with these materials, water leakage into the ducts, particularly if impurities are present, could cause serious corrosion, leading to failure of the load-carrying tendons. It was the leakage of water into the ducts combined with other unusual circumstances that produced stress corrosion cracking failures of prestressing tendons in a small-scale model of a concrete pressure vessel at Oak Ridge National Laboratory [*1*][2].

The present investigation was undertaken to determine the corrosion behavior of a typical prestressing steel in several environments and to determine the protection afforded by two different organic materials and by a portland cement grout. Generally, the corrosive environments tested were more severe than those expected to occur in any PCPV. In addition to the experimental program, the few reported incidents of prestressing steel failures in PCPV's for containment of nuclear reactors are briefly reviewed.

Experience with Prestressing Steels in Nuclear Power Stations

Prestressed concrete was first used for nuclear pressure vessels in 1960 [*2*]. As of 1976, fifteen nuclear reactor concrete pressure vessels were scheduled for operation in Europe and the United States with additional PCPV's in various stages of design and construction [*3*]. Except for the Fort St. Vrain high-temperature gas-cooled reactor where a PCPV provides primary containment, concrete vessels are used only for secondary containment in water-cooled reactors in the United States. In both cases large numbers of tendons are used; for example, in constructing the Fort St. Vrain PCPV, 448 tendons, each with 169 wires 6.4 mm (0.25 in.) in diameter, were used [*4*], and in 800- to 1100-MW(e) water-cooled reactors up to 1.1×10^6 kg (1200 tons) of prestressing steel are employed [*5*]. Despite the large number of tendons in use, incidents involving corrosion failures are extremely limited.

In 1962 and 1963, approximately five years after installation of the longitudinal and transverse cables in the containment vessels of the Marcoule G2 and G3 reactors in France, it was noted during a periodic inspection that the initial tension in some of the cables had decreased by 30 percent [*6*]. Upon unloading and examining one of these cables it was found that approximately 50 percent of the wires had broken in brittle fracture with corrosion

[2]The italic numbers in brackets refer to the list of references appended to this paper.

the apparent cause. Some parts of the circumferential cable, which was coated with several layers of a bituminous material, exhibited minor rust. Prior to examination, cable protection was provided by periodic sweeps of dried air to maintain the relative humidity of the air adjacent to the cables at 30 to 40 percent. Failure of the tendons was apparently due to hydrogen embrittlement caused by the excess humidity in the conduits. To prevent further corrosion, the relative humidity of the air was decreased to 10 percent and the airflow was changed from periodic to continuous. Since then two additional cables have been replaced, but corrosion has apparently been arrested.

Approximately eight years after initiation of construction, a statutory inspection of prestressing tendons at Wylfa Power Station in the United Kingdom in 1971 revealed extensive pitting of exposed hoop tendons with some pits as deep as 0.3 mm (0.01 in.) [*7*]. Corrosion protection of the tendons was by a grease containing a proprietary corrosion inhibitor. Laboratory studies to identify the cause of the pitting concluded that pitting was due to the combined action of contaminating salt from the sea and moisture from the air. The corrosion inhibitor prevented corrosion only up to a threshold concentration of chloride, at which point attack occurred at susceptible localized areas. The corrosion rate was also dependent on the relative humidity, with corrosion occurring only at relative humidities above 33 percent.

Despite the application of phosphate paints and heavy greases, a number of tendons installed in the bottom cap of the prestressed pressure vessel under construction at Dungeness "B" Power Station in Kent, United Kingdom, corroded badly after approximately nine months of unstressed storage in conduits [*8*]. It was later found that water had entered the tendon ducts and emulsified the grease protecting the tendon wires. Ten tendons containing 1630 wires, each 7 mm (0.28 in.) in diameter, were removed and examined, and 1550 wires were found to be severely pitted. An investigation indicated that the cause of the pitting was electrolytic attack due to an impressed anodic current from d-c welding equipment that had been grounded to metal components of the vessel. Since then, d-c welding has been prohibited in the vicinity of prestressing materials, and subsequent examinations at Dungeness have revealed no further abnormal corrosion.

The condition and functional integrity of nongrouted tendons in secondary containment structures in the United States are assessed periodically. Five vessels prestressed with tendons containing 90 wires of 6.35-mm (0.25 in.) diameter have been examined [*8*]. Of 7600 wires inspected, seven discontinuities were found with all breaks being discovered prior to the completion of construction. A total of 86 wires was removed and only three of these showed evidence of metal loss (pitting), amounting to a cross-sectional area reduction of 1 percent. This attack occurred before application of the final corrosion protection. Metallurgical examination revealed no evidence of stress corrosion cracking or hydrogen embrittlement.

Corrosion observed in the prestressing steel tendons of PCPV's which provide containment for nuclear reactors has not resulted in serious failures to this time. In retrospect, most of the corrosion damage found probably could have been prevented if proper storage, handling, and construction practices had been followed and well-established corrosion prevention practices had been employed.

Experimental Procedures

This study involved two separate phases. In one the corrosion of an uncoated typical tendon steel was investigated in several environments, and in the second the behavior of the steel coated with either organic or cement grout coatings was examined in the same corrosive environments. The high-strength AISI 1080 carbon steel selected for study was from different heats and conformed to ASTM Specifications for Uncoated Seven-Wire Stress-Relieved Strand for Prestressed Concrete (A 416-74, Grade 270). The composition of the steel is given in Table 1. Its minimum UTS is 1860 MPa (270 000 psi). The individual tendon wires were austenitized at approximately 815°C (1500°F), cooled in lead to about 500°C (930°F), and transformed at this temperature to lower pearlite. This pearlite structure was subsequently cold-worked to provide the high strength.

Corrosion Testing with Uncoated Wires

A major concern from the corrosion standpoint was stress corrosion cracking. With the uncoated steel, stress corrosion cracking tests were conducted using the constant-strain-rate method developed by Humphries and Parkins [*9*]. With this technique, a tension-type specimen while exposed to the test environment is strained at a very slow constant rate until fracture occurs, usually within a period of a few days. By comparing reduction in area with elongation or time to failure, or both, under different conditions, the relative

TABLE 1—*Composition range of ASTM A 416 Grade 270 steel used in tests.*

Element	Weight %[a]
Carbon	0.75 to 0.81
Silicon	0.26 to 0.28
Manganese	0.62 to 0.84
Phosphorus	0.012 to 0.021
Sulfur	0.018 to 0.028
Copper	0.01 to 0.02
Iron	remainder

[a]Material supplied from more than one heat.

susceptibility of a material to cracking in different environments can be established. Examination of the fracture surface and metallographic examination of the gage section near the fracture can provide additional information about the fracture process.

For the stress-corrosion cracking tests, the center straight 5-mm-diameter (0.20 in.) wire of a seven-wire tendon was used. Light surface rust was removed and the specimens were degreased in acetone before use. A gage section was not machined in the wires since we wanted to retain the original surface condition of the cold-drawn wire. The wire in the presence of the test environment was strained at a rate of 4.2×10^{-7}/s in an Instron tension machine. The length between the grips was 200 mm (8 in.) but only the center 75-mm (3 in.) length was exposed to the solution, which was held in a polyvinyl chloride bottle. To prevent any unusual effects at the solution-air interface or in the crevice where the wire passed through a rubber stopper in the bottom of the bottle, all the wire except the exposed length was coated with a stop-off varnish. After fracture, one end of the specimen was mounted axially and metallographically polished to look for secondary cracks.

The corrosion rates of unstressed wires were determined in several different aqueous environments. In one series of tests, weighed 25-mm (1 in.) lengths of wire were totally immersed in different solutions. In these tests the 5-mm-diameter (0.2 in.) wires were placed in 6-mm-diameter (0.24 in.) glass test tubes and these were covered to a depth of about 10 mm (0.4 in.) with the test solutions. The test solutions included dilute solutions of chloride (0.01 and 0.03 *M*), nitrate (0.001 and 0.01 *M*), and sulfate (0.002, 0.02, and 0.2 *M*), and potable and distilled water. The total volume of solution was only about 500 mm^3 (0.030 $in.^3$). These conditions were intended to simulate those that could exist in the interstices of wires in the tendons. Evaporative losses were replaced periodically with distilled water. Although the containers were open to the air, the small clearance between the container wall and the specimen restricted access of oxygen to the specimen. As corrosion proceeded, the buildup of corrosion products on the specimen further limited access of oxygen. These tests lasted for 2000 h, but duplicate specimens were removed from test after 1000 h. All specimens were descaled in Clarke's solution [*10*] (2-g Sb_2O_3 and 5-g $SnCl_2$ dissolved in 100 ml of 37 percent hydrochloric acid before final weighing.

In another series of tests, single wires 300 mm (12 in.) long were immersed in 0.1-*M* Na_2SO_4 and 0.05-*M* sodium chloride (NaCl) and in distilled water at 25°C (77°F) for 6500 h. The volume of solution was about 1 litre (0.26 gal) which was freely exposed to air; water lost by evaporation was replaced every few days. The test specimens extended above the water level so that attack at the water-air interface could be examined. At the end of the test the specimens were descaled, and the diameter of the wire above and below the waterline was compared.

Testing with Coated Wires

To evaluate the effectiveness of coating materials in preventing corrosion of tendon steels, two different types of tests were conducted. In one case, the specimens after coating were stressed to 60 percent of the UTS while in contact with the corrosive solution, and in the second the coated specimens were first exposed to the solution for relatively long times without an applied stress and were subsequently strained to failure. In both cases the center straight wire from the tendon was used. This wire had a diameter of 4.36 mm (0.17 in.), which was slightly less than that of the wires used in the previously described corrosion tests. The two organic coatings used in these tests, designated "A" and "B," are commercially available petroleum-based greases containing corrosion inhibitors and polar agents to facilitate wetting of the wires and displacement of moisture. They are applied according to the manufacturers' recommendations to a thickness of about 1 mm (0.04 in.). Test specimens protected by grout were prepared by casting 16-mm-diameter (0.63 in.) cylinders 38 mm (1.5 in.) long around the wires. The area of the wires not coated with a grout or grease that could be contacted with the test solutions was covered with a polyurethane insulating paint. All tests were conducted at ambient temperature except for those using ammonium nitrate solutions, and a conventional tension machine was used to stress the specimens. In some cases the effect of flaws in the coatings was evaluated. Flaws were placed in the organic coatings by scraping the wire to which the coating had been applied with a similar wire so that a thin strip of unprotected wire resulted. Flaws were cast in grout ranging in width from 0.01 to 3.2 mm (0.0004 to 0.126 in.) by using plastic shim stock which was removed after the grout hardened.

With the specimens stressed during exposure, only 0.1 *M* H_2S (pH ~ 4) was used as test solution. The organic coatings were applied to completely cover the wire surface over the 50.8-mm (2 in.) test length and the specimen was mounted in a plastic bottle as described in the preceding section. The grout-coated specimens were similarly mounted. The bottle was then filled with demineralized water and H_2S was bubbled through the water for about 15 min. The bottle was sealed and the specimen was loaded to 60 percent of its UTS and maintained at that level for six days unless the specimen failed before then. The length between the grips was 216 mm (8.5 in.). If failure did not occur within six days, the specimen was removed from the environment and pulled to failure at a crosshead velocity of 0.51 mm/min (0.02 in./min) to determine if the exposure reduced the UTS and ductility (time to failure) compared with control specimens exposed only to air.

Coated specimens in the unstressed condition were exposed to solutions of 0.1-*M* H_2S, 0.1-*M* NaCl, and 0.2-*M* NH_4NO_3. The first two were used at room temperature, but because of the stress corrosion cracking results (see later section) the ammonium nitrate solution was maintained at 66°C (150°F). A 63.5-mm-diameter (2.5 in.) polyvinyl chloride pipe with stoppers in each end was used to expose the gage length of specimens to the H_2S solution. Holes were drilled through the pipe and the specimens were centered in them by appropriately sized rubber stoppers (Fig. 1). Water in the pipe was resaturated with H_2S at frequent intervals. Exposure to the other two solutions was carried out in stainless steel pans with holes drilled through the bottoms, and the specimens were held in position by means of rubber stoppers. After various exposure times, specimens were removed from the test environment and pulled to failure at a crosshead speed of 0.51 mm/min (0.02 in./min).

Results

Stress Corrosion Cracking Tests

The results from the slow constant-strain-rate tests are given in Table 2. Test 1 confirmed that the UTS of the steel met the requirements of ASTM A 416-74, Grade 270. Tests 2 through 6 were conducted with 0.2-*M* NH_4NO_3 since this reagent was suspected of causing failure in the thermal cylinder test at Oak Ridge National Laboratory [*1*]. No evidence of cracks was found at 21 and 38°C (70 and 100°F), but cracks formed at 52 and 66°C (126 and 151°F). Figure 2 shows representative views along the axis of the specimens tested at the four temperatures. Even though cracks were evident in the steel at the two higher temperatures, no effect on either time to failure or on the load at failure was apparent. The large diameter of the wire and the slow rate of crack propagation were probably responsible for this observation. The formation of cracks at 66°C (151°F) and the absence of cracks at 21°C (70°F) agree with the results from U-bend specimens exposed to the same solution; U-bends cracked in a few days at 66°C (151°F) but remained intact at 21°C (70°F) during 100-day tests [*1*].

Cracks were not found in any of the specimens exposed to chloride-containing solutions (Tests 7 to 10, Table 2) regardless of the pH of the solution. These results are in conformance with the generally accepted belief that chloride ions do not produce cracking in steels of this type.

As expected, H_2S produced rapid failure in the test specimens when the pH was low and the H_2S concentration was relatively high (Tests 11 and 12, Table 2). Figure 3 shows cracks observed in Test 12. In 0.001-*M* Na_2S at a pH of 11.1 or 7.4 (Tests 13 and 14), no cracking was observed, but, when the pH was adjusted to 4.3 (Test 15), cracking occurred. At comparable pH values, failure took longer at the lower sulfide concentration than at the higher concentration. Cracking occurred slightly faster when the Na_2S concentration was increased to 0.003 *M* (Test 16). When the sulfide concentration was reduced to 0.0003 *M* (Tests 17 and 18), no cracking was observed

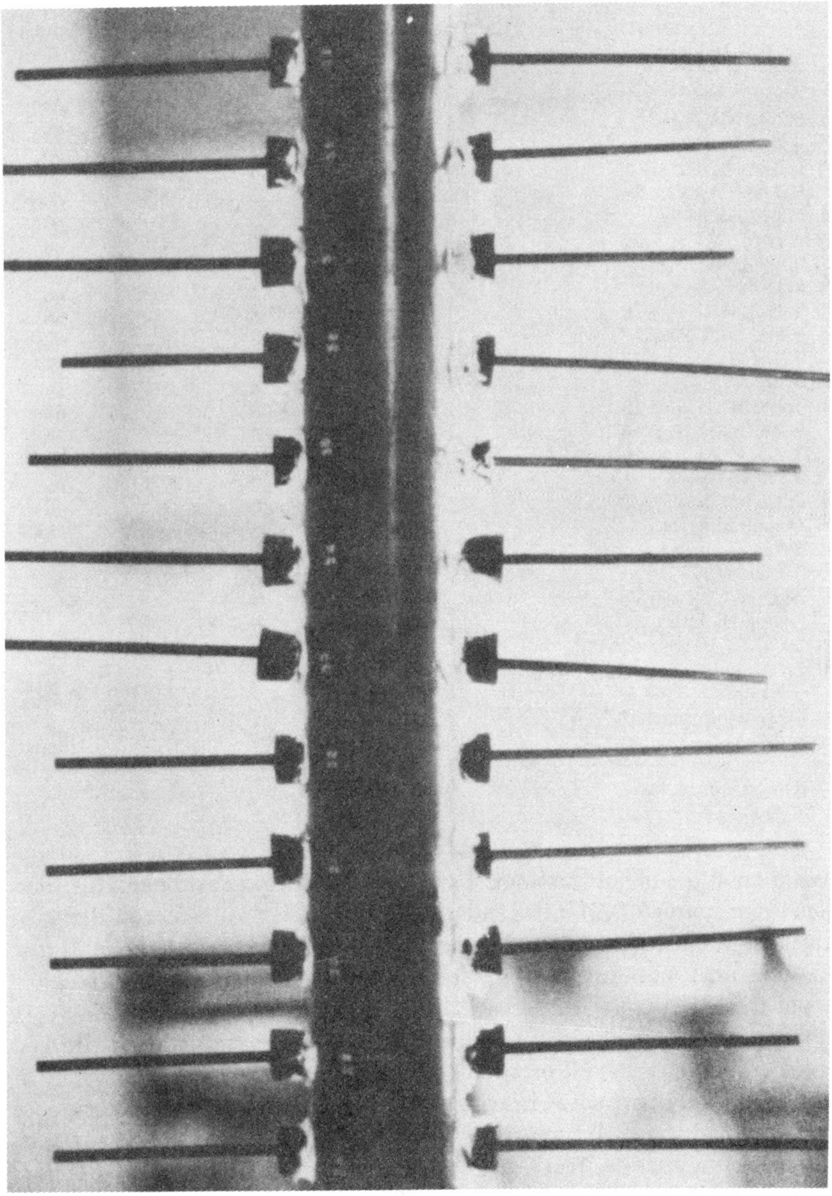

FIG. 1—*Photograph of test specimens mounted in polyvinyl chloride pipe.*

TABLE 2—*Results obtained from constant-strain-rate tests with high-strength steel wires (strain rate = 4.2 × 10^−7/s).*

Test Medium	Temperature, °C	pH	Time to break, h	Load, kN[a]	Cracks
1. Air	21[b]		55.5	41.9	no
2. 0.2-*M* NH_4NO_3	21	5.0	57.5	40.1	no
3. 0.2-*M* NH_4NO_3	21	5.3	66.5	39.3	no
4. 0.2-*M* NH_4NO_3	38	5.3	47.0	39.0	no
5. 0.2-*M* NH_4NO_3	52	5.3	49.5	38.9	yes
6. 0.2-*M* NH_4NO_3	66	5.0	50.5	38.9	yes
7. 0.01-*M* KCl + CaO	21	11.9	55.3	39.1	no
8. 0.01-*M* KCl	66	5.3	60.8	38.9	no
9. 0.03-*M* KCl	21	5.5	58.1	39.3	no
10. 0.03-*M* KCl + HCl	21	3.0	60.8	39.3	no
11. 0.1-*M* H_2S + HCl	21	3.0	6.6	15.0	yes
12. 0.1-*M* H_2S	21	4	4.7	11.1	yes
13. 0.001-*M* Na_2S	21	11.1	57.8	38.4	no
14. 0.001-*M* Na_2S + HCl	21	7.4	53.8	39.1	no
15. 0.001-*M* Na_2S + HCl	21	4.3	19.3	35.2	yes
16. 0.003-*M* Na_2S + HCl	21	4.3	17.0	34.2	yes
17. 0.0003-*M* Na_2S + HCl	21	6.3	52.0	39.3	no
18. 0.0003-*M* Na_2S + HCl	21	4.3	54.8	39.3	no
19. Deionized water	21	6.0	57.5	39.0	no
20. Deionized water + CO_2	21	4	55.8	39.3	no
21. H_2O + CO_2 + 750 ppm H_3AsO_3	21	5.1	57.6	39.3	no
22. Zn plate, H_2O	21	5.8	80.6	39.5	no
23. Zn plate, 0.2-*M* Na_2SO_4	21	5.7	54.9	39.5	no
24. Corroded in water, 3 days	21	5.9	50.9	39.3	no

[a]lb = Newton/4.448.
[b]°F = (°C × 9/5) + 32.

even though the pH of the solutions was 4.3 in one case. Although the solutions were sparged with nitrogen before the Na_2S was added, and the polyvinyl chloride bottle containing the solution was stoppered, it is probable that the small amount of sulfide in the solution was oxidized to sulfur early in the test by traces of oxygen in the bottle.

The other data in Table 2 show that cracking was not produced in distilled water with carbon dioxide or when arsenious acid was added to the water. In the latter case the wire was heavily pickled in uninhibited hydrochloric acid and remained in the test solution overnight before the specimen was pulled. Electroplating the specimen with zinc, except for an area 200 mm^2 (0.3 $in.^2$) in the middle of the wire, also had no effect on cracking in either water or a dilute sodium sulfate solution.

In all cases except where cracking occurred in the presence of sulfide, cup-cone-type failures were observed, and the reduction in area varied randomly

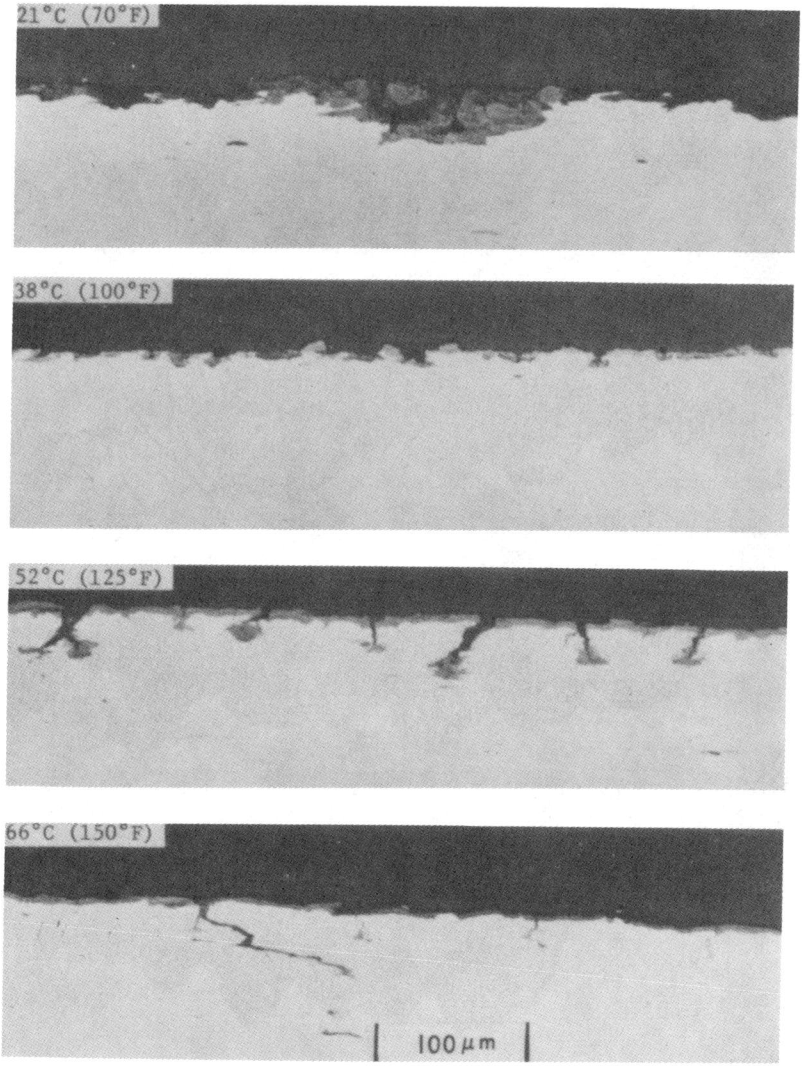

FIG. 2—*Attack on steel tendons wires in 0.2-*M NH_4NO_3. *Strain rate,* 4.2×10^{-7}/s.

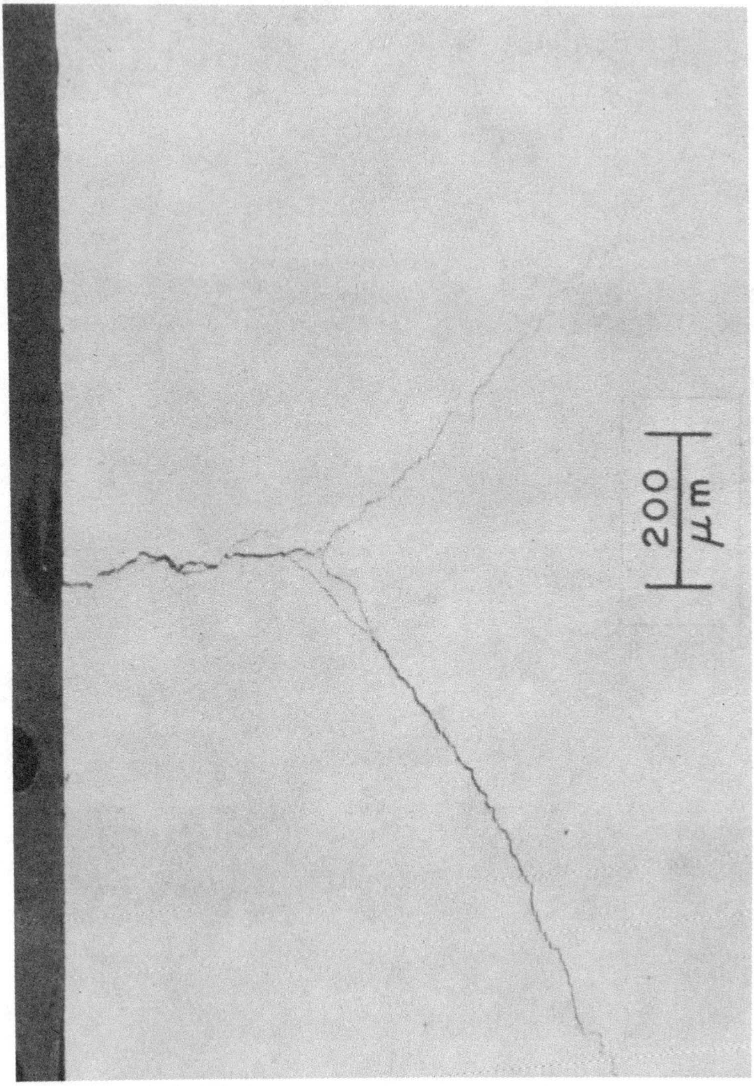

FIG. 3—*Cracks in tendon wire formed during constant-strain-rate test in 0.1-M* H_2S.

between 36 and 43 percent except in Test 2, where it was 47 percent. In Tests 11 and 12, no reduction in area could be measured, and in Tests 15 and 16 the reduction in area was only 4 percent. Even though cracks were not observed in most cases, most of the failures occurred in the 75-mm (3 in.) length exposed to the test solution.

General Corrosion Rate Determinations

The corrosion rates observed for the tendon wires exposed to very small volumes of solutions are given in Table 3. Each value represents a single specimen. After 1000 h in some cases and after 2000 h in most cases, the volume of corrosion products prevented removal of the specimen from the glass container, and it was necessary to break the container to recover the specimen. Table 3 shows that, under these conditions, corrosion rates were low and nearly independent of the environment. There was no significant localized attack in any case. Vacuum fusion analysis of specimens exposed to each solution for 2000 h showed hydrogen content increases ranging from less than 1 ppm to a maximum of 2.4 ppm.

The 300-mm-long (12 in.) wires exposed to large volumes of solution that was freely exposed to air corroded more than the specimens with limited air exposure. Corrosion rates calculated from differences in diameter above and below the waterline after 6500 h were 76, 152, and 254 μm/year (3, 6, 10 mils/year), respectively, for distilled water, 0.05-*M* NaCl, and 0.1-*M* Na_2SO_4. Attack was greatest at the solution-air interface, but localized attack in the form of wide elongated pits was noted below the waterline in all cases. Therefore, maximum penetration rates were considerably greater than

TABLE 3—*Corrosion of tendon wires in different environments with restricted access to oxygen.*[a]

	Corrosion Rate, μm/year[b]	
Test Environment	1000 h	2000 h
Potable water	5.8	5.3
Distilled water	5.1	5.3
0.01-*M* NaCl	6.4	9.7
0.03-*M* NaCl	6.6	7.9
0.001-*M* $NaNO_3$	7.4	23.1
0.01-*M* $NaNO_3$	11.2	8.1
0.002-*M* Na_2SO_4	5.8	7.6
0.02-*M* Na_2SO_4	7.6	13.2
0.2-*M* Na_2SO_4	8.9	13.5

[a]The pH of all solutions ranged between 5.5 and 6.5.
[b]1 mil/year = 25.4 μm/year.

indicated in the foregoing. Comparison of these corrosion rates with those shown in Table 3 clearly shows that free access to oxygen greatly accelerates corrosion.

Coated Wires Stressed During Exposure

A series of tests with bare wires exposed to 0.1-*M* H_2S showed that, as the stress level increased, both the time to failure and the scatter in the data decreased up to about 50 percent of the failure stress (Fig. 4). From 50 to 90 percent, failures occurred in 1 to 2 h, independent of stress level.

Results obtained with the coated specimens that were stressed during exposure to 0.1-*M* H_2S are summarized in Table 4. None of the specimens covered with unflawed coatings of either of the organic greases or grout cracked during approximately 6-day exposures at 60 percent UTS, and subsequent tension tests showed that no degradation of either load-carrying capacity or ductility (time to failure) had occurred. Certain-sized flaws in either organic coating resulted in fracture of the wires in short times. Flaw widths of 0.1 and 0.3 mm (0.004 and 0.012 in.) in the grout produced no detrimental effects, but as the flaw width increased beyond 0.3 mm (0.012 in.) cracking failures at 60 percent UTS occurred in progressively shorter times.

Coated Wires Unstressed During Exposure

Tables 5, 6, and 7 summarize the results obtained with both coated and uncoated wires that were exposed for various times to 0.1-*M* H_2S, 0.2-*M* NH_4NO_3, and 0.1-*M* NaCl, respectively, without applied stress and then subsequently tension tested. The average failure load and time to failure for unexposed specimens were 28.62 kN (6430 lb) and 34.65 min, respectively.

Table 5 shows that for wires completely covered with either organic material or cement grout, exposure to H_2S-saturated water for up to 120 days had little or no effect on UTS or ductility. Completely unprotected specimens under the same condition showed a minor decrease in UTS and a major loss of ductility. Specimens with flaws in the organic coatings showed significant losses in both ductility and UTS after 119 days and proportionately smaller losses at shorter times (not shown in Table 5). On the other hand, flaws up to 0.76 mm (0.03 in.) wide in portland cement grout produced only very minor effects even after 119 days.

In the ammonium nitrate solutions (Table 6) the unprotected wires lost about 14 percent in UTS and 66 percent in ductility after 42 days and no further changes were noted for the remainder of the 132-day test. Both organic materials provided complete protection for the duration of the test, but the protectiveness of the grout seemed to decrease slowly with time. The am-

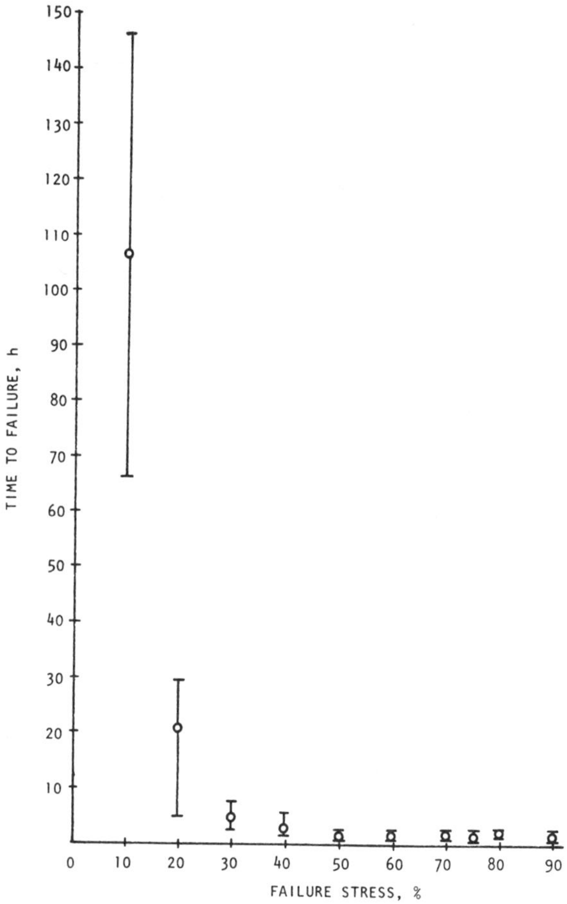

FIG. 4—*Effect of stress (percent UTS) on time to failure for tendon wire in 0.1-*M H_2S.

monium nitrate solution at 66°C (151°F) causes degradation of grout, a fact which probably allowed ammonium nitrate to come in direct contact with the steel after some time. Specimens with flaws in the grout showed about the same result. On the other hand, specimens with flaws in the organic coatings showed little if any loss in properties, possibly because at 66°C (151°F) the organic greases flowed over the defected areas early in the test and protected them.

In the 0.1-*M* NaCl solution (Table 7), wires protected by all three materials seemed to slowly decrease in UTS and time to failure so that after 164 days the loss was 5 to 7 percent in both cases. Unprotected specimens behaved in a similar fashion but the percentage losses in UTS and time to failure were 13 and 64 percent, respectively, after 164 days. Specimens in which the organic

TABLE 4—*Loads at failure and failure times for specimens exposed in 0.1-*M H_2S *at room temperature.*

Coating Material	Average Failure Load,[a] kN[c]	Average Failure Time, h
Organic "A"		
unflawed	28.58	152.0[b]
flawed	17.17	14.0
Organic "B"		
unflawed	28.63	149.0[b]
flawed	17.17	63.8
Portland cement grout		
unflawed	28.36	162.7[b]
0.1-mm[d] flaw	28.60	260.0[b]
0.3-mm flaw	27.86	180.9[b]
1.3-mm flaw	17.17	118.3
1.6-mm flaw	17.17	48.1
3.2-mm flaw	17.17	34.6

[a]Average load and time to failure for control specimens were 28.62 kN and 34.7 min, respectively.
[b]No failure occurred for indicated exposure at 60 percent UTS; failure load obtained from standard tension test.
[c]lb = Newton/4.448.
[d]mm = 0.04 in.

TABLE 5—*Average load at failure and time to failure for wires exposed to 0.1-*M H_2S *without applied stress and subsequently pulled to failure.*

Coating Material	Exposure Time, days	Average Load at Failure,[a] kN[b]	Average Time to Failure,[a] min
None	33	26.28	13.2
Organic "A"	33	27.84	34.2
Organic "B"	33	27.98	32.6
Portland cement grout	33	27.95	32.5
None	77	26.37	13.8
Organic "A"	77	27.65	31.9
Organic "B"	77	28.06	35.9
Portland cement grout	77	28.02	32.9
None	120	25.36	10.4
Organic "A"	120	...	...
Organic "B"	120	28.65	29.8
Portland cement grout	120	28.43	30.6
Flawed organic "A"	119	23.91	10.5
Flawed organic "B"	119	24.24	12.5
Portland cement grout			
0.01-mm[c] flaw	119	26.71	33.3
0.10-mm flaw	119	26.82	34.8
0.51-mm flaw	119	26.13	32.8
0.76-mm flaw	119	25.58	32.0

[a]Average load and time to failure for control specimens were 28.62 kN and 34.7 min, respectively.
[b]lb = Newton/4.448.
[c]mm = 0.04 in.

TABLE 6—*Average load at failure and time to failure for wires exposed to 0.2-M NH_4NO_3 at 66°C (151°F) without applied stress and subsequently pulled to failure.*

Coating Material	Exposure Time, days	Average Load at Failure,[a] kN[b]	Average Time to Failure,[a] min
None	42	24.58	11.5
Organic "A"	42	27.32	36.2
Organic "B"	42	27.24	35.1
Portland cement grout	38	28.62	34.7
None	78	26.77	14.7
Organic "A"	78	28.36	34.1
Organic "B"	78	38.39	37.1
Portland cement grout	87	38.14	27.2
None	132	25.91	11.7
Organic "A"	132	28.65	35.6
Organic "B"	132	28.47	34.3
Portland cement grout	122	26.47	25.2
Flawed organic "A"	130	26.47	36.0
Flawed "B"	130	26.80	33.3
Portland cement grout			
0.03-mm[c] flaw	130	26.36	25.0
0.25-mm flaw	130	25.80	23.0
0.76-mm flaw	130	25.69	22.3
3.18-mm flaw	130	25.91	19.3

[a]Average load and time to failure for control specimens were 28.62 kN and 34.7 min, respectively.
[b]lb = Newton/4.448.
[c]mm = 0.04 in.

coatings contained flaws showed about the same behavior as the unflawed materials. Flaws from 0.03 to 0.76 mm (0.0012 to 0.03 in.) wide in grout coatings had no significant effect, but for a flaw 3.18 mm (0.127 in.) wide the mechanical properties of the steel were slightly less after 153 days than totally unprotected specimens exposed for 164 days.

Discussion

The steel wires used in this study were made from AISI 1080 steel which was tempered and cold-drawn to obtain the high strength required of tendon steels. At room temperature this steel was not susceptible to conventional stress corrosion cracking in the presence of impurities usually found in grouts or organic coatings (sulfates, nitrates, chlorides), but under some conditions this steel was subject to hydrogen embrittlement. Cracks were produced in ammonium nitrate solution at temperatures above room temperature. Small water pockets that could exist within grouts or organic materials containing alkaline-producing additives would have high pH values, and it is doubtful that cracking would occur in the presence of ammonium nitrate under such conditions. It should be noted, however, that even in alkaline grouts, calcium chloride (2 to 5 percent) produces severe pitting of embedded steel [*11,12*].

TABLE 7—*Average load at failure and time to failure for wires exposed to 0.1-M NaCl without applied stress and subsequently pulled to failure.*

Coating Material	Exposure Time, days	Average Load at Failure,[a] kN[b]	Average Time to Failure,[a] min
None	71	27.53	17.8
Organic "A"	71	28.62	36.1
Organic "B"	71	28.47	36.4
Portland cement grout	71	28.51	37.8
None	107	25.72	13.9
Organic "A"	107	27.13	30.4
Organic "B"	107	26.99	29.5
Portland cement grout	107	27.02	32.9
None	164	24.76	12.6
Organic "A"	164	26.34	33.1
Organic "B"	164	26.43	32.8
Portland cement grout	164	26.80	32.4
Flawed organic "A"	153	25.58	34.8
Flawed organic "B"	153	25.13	30.3
Portland cement grout			
0.03-mm[c] flaw	153	26.58	31.5
0.25-mm flaw	153	26.91	36.5
0.76-mm flaw	153	26.24	37.3
3.18-mm flaw	153	23.58	8.8

[a]Average load and time to failure for control specimens were 28.62 kN and 34.7 min, respectively.
[b]lb = Newton/4.448.
[c]mm = 0.04 in.

Test wires failed in short times when strained in solutions of H_2S in water at a pH of 4 or less, but where the pH was raised to 7 or above, cracking failures did not occur. There are, however, reported incidents of cracking occurring at pH values as high [*13,14*] as 9.5. Cracking of high-strength steels in the presence of H_2S has been attributed to hydrogen embrittlement, the H_2S facilitating the entry of hydrogen into the steel by interfering with the formation of molecular hydrogen at cathodic areas [*15*]. Only molecular hydrogen sulfide is effective in causing hydrogen embrittlement [*13,14*], and, as the pH of a solution is increased, H_2S dissociates into HS^- and $S^=$ ions and cracking does not occur. At the pH of correctly formulated concrete (>12), small amounts of H_2S in the cement do not effect the mechanical properties of the embedded steel. A similar situation should exist with steel coated with greases that contain alkaline-producing additives.

The corrosiveness of several dilute neutral solutions to bare high-strength steel wires depended on the availability of oxygen. With restricted access to oxygen, corrosion rates were significantly lower than when the solutions were freely exposed to air. Although we did not investigate the effect of pH on the corrosion rate of these steels, it is likely that even in the presence of oxygen the corrosion rates would be very low in alkaline solutions. Therefore, if pockets of water in contact with tendons should develop in grout- or grease-filled conduits, corrosion of the steel would be low because of limited availability of oxygen and because of the alkalinity of the solution in contact with grouts or organic materials containing alkaline additives. In such cases, hydrogen embrittlement produced by the corrosion reaction is unlikely since even in neutral solutions very little hydrogen was picked up by the steel in 2000-h tests.

The three coatings applied to the test specimens provided almost complete protection even though they were tested in environments much more aggressive than would be expected in a PCPV. Those specimens coated with either of the two organic materials selected for study showed no signs of attack on exposure to 0.1-M H_2S when stressed at 60 percent UTS for about 6 days. Subsequent tension tests showed no degradation of mechanical properties. If a small amount of coating was removed, however, failures occurred in relatively short times. Portland cement in the same type of test also provided complete protection, and with flaws up to 0.3 mm (0.012 in.) wide no failures occurred; however, with flaw widths of 1.3 mm (0.052 in.) or greater, failures took place in less than 6 days. It seems probable that with a sufficiently small flaw or crack in the relatively thick concrete coating, the pH of the small volume of solution within the crack remained high enough so that the H_2S dissociated and cracks could not form.

Exposure of unstressed coated specimens to 0.1-M H_2S and tension testing also showed the coatings to be completely effective, if unflawed. However, exposure of uncoated specimens and organic coatings with flaws to 0.1-M H_2S caused slight losses of load-carrying capacity and substantial reductions in times to failure during tension tests. On the other hand, concrete coatings with flaws as wide as 0.76 mm (0.03 in.) resulted in only very small losses in mechanical properties after exposure to H_2S. Similar results were obtained in ammonium nitrate and sodium chloride solutions. In the ammonium nitrate solution, flaws in either organic coating did not cause loss of mechanical properties, apparently because at the test temperature of 66°C (151°F) the organic materials flowed over the flaws and protected them.

Summary

The stress-corrosion cracking susceptibility of a typical cold-drawn high-strength steel (ASTM A 416-74 steel, Grade 270) to several solutions was determined using a slow constant-strain-rate technique. The steel developed cracks in 0.2-M NH_4NO_3 but only at temperatures above 38°C (100°F). Brittle failures, apparently because of hydrogen embrittlement, also occurred in hydrogen sulfide solutions if the pH was less than 7. Cracking did not occur in chloride solutions regardless of pH. The corrosion rate of the steel was low in dilute solutions of NaCl, $NaNO_3$, and Na_2SO_4 when access to oxygen was restricted, but it was substantially higher with free access to air; in the latter case, broad pits formed, even in pure water.

Both organic coating materials tested and portland cement grout provided complete protection to the steel in the aforementioned aggressive environment, provided the coating remained intact. With portland cement, crack widths up to 0.76 mm (0.03 in.) did not result in loss of protection. These results indicate that in a PCPV the use of certain commercial organic greases or portland cement grout for filling the conduits containing high-strength steel tendons should provide protection to the tendons if low concentrations of aggressive salts inadvertently get into the conduits during filling.

Acknowledgment

This research was sponsored jointly by the U.S. Nuclear Regulatory Commission under Interagency Agreements DOE 40-551-75 and 40-552-75 and the Nuclear Power Development Division of the U.S. Department of Energy under Contract W-7405-eng-26 with the Union Carbide Corp.

References

[*1*] Canonico, D. A., Griess, J. C., and Robinson, G. C., "Final Report on PCRV Thermal Cylinder Axial Tendon Failures," ORNL-5110, Oak Ridge National Laboratory, Oak Ridge, Tenn., Jan. 1976.

[*2*] Germain, F., *Journal of the Prestressed Concrete Institute*, Vol. 12, No. 4, Aug. 1967, pp. 43-52.

[*3*] Hannah, I. W., in *Proceedings*, Symposium on Experience in the Design, Construction, and Operation of Prestressed Concrete Pressure Vessels and Containments for Nuclear Reactors, Paper 150/75, University of York, England, 8-12, Sept. 1975.

[*4*] Hildebrand, J. F. in *Proceedings*, Symposium on Experience in the Design, Construction and Operation of Prestressed Concrete Pressure Vessel, and Containments for Nuclear Reactors, Paper 150/75, University of York, England, 8-12 Sept. 1975.

[*5*] Schupack, M, *Journal of the Prestressed Concrete Institute*, Vol. 17, No. 3, May/June, 1972, pp. 14-28.

[*6*] Poitevin, P., "Containers of the G2 and G3 Reactors: Protection of Prestressing Cables," Technical Information Note Campagnie Industrielle de Travaux (Engerprises Schneider), ORNL translation STS 5651, Oak Ridge National Laboratory, Oak Ridge, Tenn., 21 March, 1966.

[*7*] Fontain, M. J. et al in *Proceedings*, Symposium on Experience in Design, Construction, and Operation of Prestressed Concrete Pressure Vessels and Containments for Nuclear Reactors, Paper 149/75 University of York, England 8-12 Sept. 1975.

[*8*] Rotz, J. V. in *Proceedings*, Symposium on Experience in Design, Construction, and Operation of Prestressed Concrete Pressure Vessels and Containments for Nuclear Reactors, Paper 172/75, University of York, England, 8-12 Sept. 1975.

[*9*] Humphries, M. J. and Parkins, R. N. in *Fundamental Aspects of Stress Corrosion Cracking*, R. W. Staehle, Ed., National Association of Corrosion Engineers, Houston, Tex., 1969, pp. 384-392.

[*10*] Clarke, S. C., *Transactions*, Electrochemical Society, Vol. 69, 1936, pp. 131-144.

[*11*] Monfore, G. E. and Verbeck, G. J., *American Concrete Institute Journal*, Vol. 57, No. 11, 1960, pp. 491-515.

[*12*] Treadaway, K. W. T., *British Corrosion Journal*, Vol. 6, 1971, pp. 66-72.

[*13*] Hudgins, C. M. et al, *Corrosion*, Vol. 22, 1966, pp. 238-251.

[*14*] McCord, T. G. et al, *Materials Performance*, Vol. 15, No. 2, 1976, pp. 25-34.

[*15*] Greer, J. B., *Materials Performance*, Vol. 14, No. 3, 1975, pp. 11-22.

S-S. Yau[1] *and W. H. Hartt*[1]

Influence of Selected Chelating Admixtures upon Concrete Cracking Due to Embedded Metal Corrosion

REFERENCE: Yau, S-S. and Hartt, W. H., "**Influence of Selected Chelating Admixtures upon Concrete Cracking Due to Embedded Metal Corrosion,**" *Corrosion of Reinforcing Steel in Concrete, ASTM STP 713,* D. E. Tonini and J. M. Gaidis, Eds., American Society for Testing and Materials, 1980, pp. 51-63.

ABSTRACT: It has been considered by the present research that if the solubility limit for embedded metal corrosion products can be increased or the products retained in solution altogether, then development of tensile stresses within the concrete and resultant cracking should be prolonged or eliminated. Consequently, the influence of selected chelating agents, including TEA, EDTA, DPTA, HEDTA, and Chel-138, upon various concrete properties has been evaluated. Specific tests investigated (1) the influence of each agent upon iron solubility as a function of pH, (2) the influence of the various agents as admixtures upon concrete strength, (3) the effect of the chelating agents upon corrosion of steel in aqueous solutions of pH 10 and 12 with 0 and 0.1 percent sodium chloride, and (4) the influence of the agents upon time-to-cracking of reinforced concrete cylinders in an accelerated test. While results of tests in Categories 1 and 3 were encouraging, it was determined that the chelating additions reduced concrete compressive strength. Further, time-to-cracking of admixtured specimens was extended beyond that of standard (no admixture) ones only for HEDTA and Chel-138 and then only under certain test conditions. It is concluded that, while the concept of adding chelating agents to concrete to enhance cracking resistance may be sound, increased time-to-cracking was not generally observed because of reactions between the various admixtures and the concrete.

KEY WORDS: reinforced concrete, admixtures, corrosion, concrete cracking, chelating agents

Cracking and spalling of concrete due to embedded metal corrosion are presently recognized both technologically and economically as a serious problem. The damage is generally attributed to breakdown of the passive film on the steel from chloride ion penetration to the metal-concrete interface and to consequent development of active-passive cells or concentration cells or both. This leads ultimately to accumulation within the concrete pores near

[1]Professor and graduate research assistant, respectively, Department of Ocean Engineering, Florida Atlantic University, Boca Raton, Fla. 33431.

the embedded metal surface of insoluble corrosion products and, subsequently, to development of tensile hoop stresses and cracking [*1,2*].[2] Variables which have been determined to be influential regarding this phenomenon include cement type [*3*], concrete mix design [*2,4*], and chloride ion concentration.

A facet of this problem that has received little attention pertains to the solubility limit of corrosion products in the vicinity of the metal-concrete interface. Thus, it may be reasoned that as long as such products remain in solution tensile stresses and cracking should not develop.

The objective of the present research was to evaluate the influence of selected admixtures, which are thought to retain iron ions in solution, upon properties of reinforced concrete and cracking tendency.

Experimental Procedure

The research plan involved four different classes of experiments, which include tests to (1) determine the influence of the selected admixtures upon iron ion solubility limit, (2) evaluate the influence of the admixtures upon the corrosion rate of steel, (3) determine the relative cracking tendency of reinforced concrete specimens with and without solubility-enhancing additions, and (4) evaluate the influence of admixtures upon concrete properties.

The admixtures were all of the iron (III) chelating variety, as listed in Table 1. Thus, the test philosophy considered that corrosion products based upon the ferric ion are important with regard to cracking.

Solubility Determinations

For the purpose of determining iron ion solubility limit as a function of pH and chelating agent concentration, an iron standard solution (1 mg/ml Fe^{3+} as $FeCl_3$ in dilute hydrochloric acid) was added to buffered solutions of various pH until visible particulates were observed. The solution in this case was assumed to be saturated with iron. Because of the acidity of the iron standard, the pH of the final solution differed from that typical of the buffer. Initial experiments indicated that an acceptable compromise between the requirement on the one hand to add significant iron and on the other to maintain a given pH was a net solution comprising 2 ml of the iron standard and 48-ml buffer. A 24-h period was allowed for any reactions to stabilize. Solutions were then filtered using Fisher 15-cm CP-QL qualitative-grade paper to remove particulates. Determination of iron concentration in the solutions was made using a Perkin-Elmer Atomic Absorption Spectrophotometer, Model 305. From sequential atomic absorption measurement and filtering it was determined that iron concentration of a solution decreased with each

[2]The italic numbers in brackets refer to the list of references appended to this paper.

TABLE 1—*Iron chelating agents investigated in the present study.*

	Chelating Agent	Comments
1.	EDTA	employed as rust-removing agent effective in holding relatively large amounts of iron in solution effectiveness decreases with increasing pH above 8
2.	DTPA	effectiveness at a given pH is reported to be less than for EDTA
3.	HEDTA	this is considered to be a more effective agent at pH's above 8
4.	TEA	this is reported to be an effective chelating agent for solutions with pH greater than 12
5.	Chel- 138	this is considered to be the most potent such material for high pH applications

filtering but was constant after three passes. Supplementary experiments employed centrifuging to remove particulates; however, results of these tests were the same as for filtering. Solution pH determinations involved an Orion digital pH/mV meter, Model 701A, in conjunction with a Markson combination electrode, Model 1207. Solution temperature in all instances was that of the laboratory air, which was approximately 23°C.

Subsequent to the preceding determinations the experiments were repeated, but with solutions containing various concentrations of chelating agent in addition to the buffer.

Steel Corrosion Properties

Determination of any influence of the various admixtures upon corrosion properties of steel specimens involved performing anodic polarization scans as a function of pH, chelating agent concentration, and chloride ion concentration. Instrumentation included a Wenking Fast Rise Potentiostat, Model 68FRO.5, a Wenking Stepping Motor Potentiometer, Model SMP 69, and a Hewlett-Packard X-Y Recorder, Model 7001 AR. Supplementary current and potential measurements were made with two Ballintine digital multimeters, Models 3026A and 3028A, respectively. The electrochemical cell was a Princeton Applied Research Corrosion Cell System, which included two graphite counterelectrodes, a saturated calomel reference electrode, a five-neck flask, and a flat trifluoroethylene resin specimen holder which exposed 1 cm^2 of metal surface to the electrolyte. Specimens were prepared from 0.025-cm AISI 1010 steel shim stock by polishing through 600-grit silicon carbide paper, rinsing with distilled water, and drying. Prior to scanning, each specimen was exposed to the electrolyte for 30 min, which was sufficient for the corrosion potential to stabilize to a relatively constant value. The scan rate was 20 mV/min, and the test was continued until the potential was 50 mV more positive than the pitting potential or the potential for oxygen evolution.

Cracking Tendency Experiments

These tests involved exposure of 10.2-cm-diameter by 15.2-cm-high reinforced concrete cylinders to flowing seawater in conjunction with an anodic direct current impressed upon the embedded metal. During the tests, the magnitude of the impressed current was maintained constant, and the time for a visible crack to appear on the external surface was determined. Details of the apparatus and procedure for these experiments have been reported previously [*6*].

Results and Discussion

Solubility Limit Experiments

Figure 1 projects the results of iron ion determinations as a function of the type of chelating agent and of pH. From this it is apparent that in the intermediate pH range the various agents in the concentrations investigated enhance iron ion solubility by approximately an order of magnitude or more. Because of the limitations of the experimental technique, however, any solubility increase for pH > 9 was less well defined. This is unfortunate in

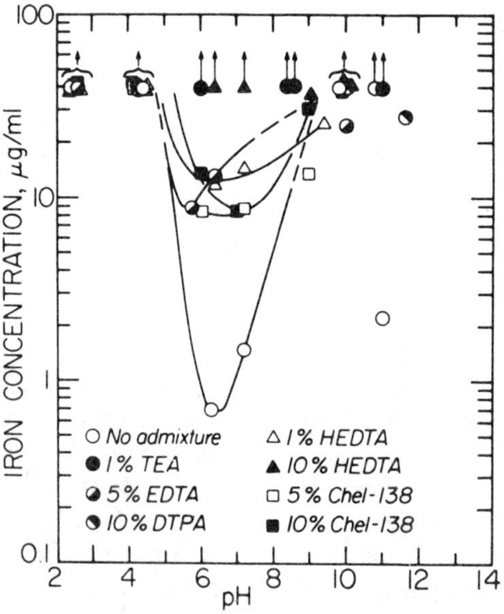

FIG. 1—*Iron solubility limit as a function of pH and type of chelating agent.*

view of the fact that the pH range 12.0 to 12.4 is thought to be important to concrete pore water [7]. However, Grimes et al [6] have shown that once corrosion occurs, pH in the vicinity of the metal-concrete interface can drop to near-neutral or acid values.

Compressive Strength Experiments

Based upon results of the various chelating agents upon iron ion solubility (Fig. 1) a concentration or concentrations of each were selected for inclusion as an admixture in concrete test specimens. Table 2 lists the standard mix design for these, and Table 3 indicates the amount of each chelating chemical which was prescribed. All specimens were prepared by the Florida Department of Transportation Office of Materials and Research, and compressive strength determinations were performed upon standard 15.2 by 30.5-cm cylinders following a 28-day moist cure. Table 4 lists results of these tests for the various mixes; and it is apparent that each of the chelating additions had an adverse effect upon strength, the decrease ranging from 3 to 40 percent in comparison with the standard (no admixture) case.

Anodic Polarization Scans

In tests to disclose the influence of chelating admixtures upon corrosion properties of steel, particular attention was focused upon HEDTA and Chel-138, because these apparently compromised concrete strength the least

TABLE 2—*Mix design for concrete specimens.*

Material	Quantity
Cement (portland Type I)	15.4 kg
Fine aggregate	23.1 kg
Coarse aggregate	46.3 kg
Water	5.7 litres

TABLE 3—*Concentration of chelating admixtures which were employed in the present concrete mix design.*

Chelating Admixture	Volume % to Water
TEA	1.0
EDTA	5.0
DTPA	1.0
HEDTA	10
Chel-138	10

(see Table 4). Figures 2 and 3 illustrate the anodic polarization behavior for steel in a pH 10 and 12 buffer solution, respectively, with 0, 1, 5, and 10 percent HEDTA. Figures 4 and 5 report results for identical concentrations of Chel-138. Figures 6–9 present data for steel in these same solutions but with addition of 0.1 percent sodium chloride (0.1NaCl).

In the former case (Figs. 2–5 with no chlorides) passive behavior was apparent with a passive current density of 2 to 4 $\mu A/cm^2$. In the case of 0.1NaCl solutions at pH 10, such behavior (passivity) was either absent or much less apparent. Passivity is more evident for the 0.1NaCl, pH 12 buffer solutions (Figs. 8 and 9) for which passive current density was variable and in the range 1 to 5 $\mu A/cm^2$. Of importance in Figs. 6–9 is the observation that the critical pitting potential was more negative with increased chelating agent

TABLE 4—*Compressive strength of concrete subsequent to 28-day moist cure.*

Admixture	Compressive Strength (28 days), MN/m^2 (ksi)
None	41.4 (6)
TEA	24.8 (3.6)
EDTA	31.7 (4.6)
DTPA	37.2 (5.4)
HEDTA	39.3 (5.7)
Chel-138	40.0 (5.8)

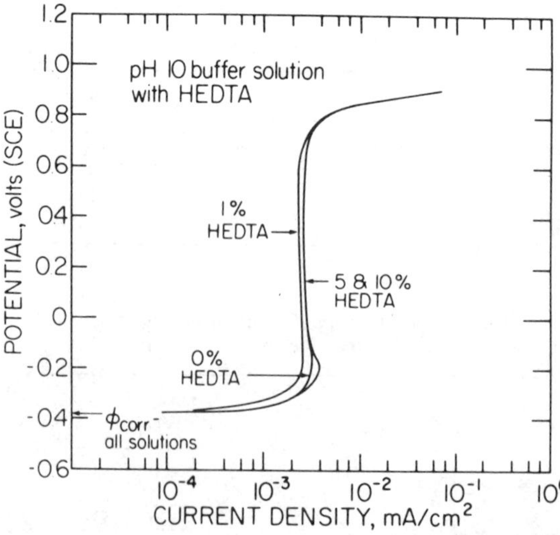

FIG. 2—*Anodic polarization curve for steel in pH 10 buffer solution as a function of volume percent HEDTA.*

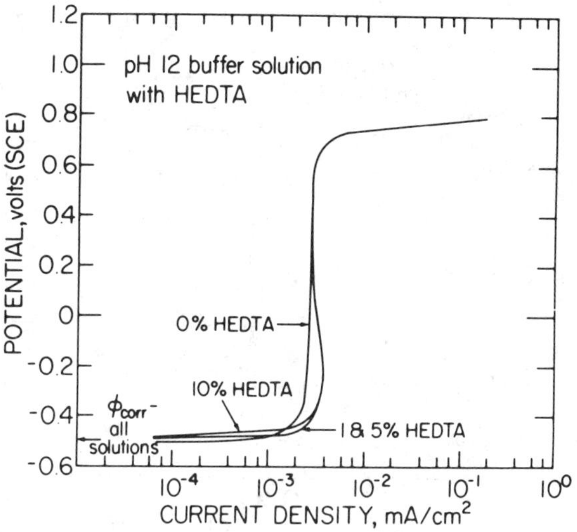

FIG. 3—*Anodic polarization curve for steel in pH 12 buffer solution as a function of volume percent HEDTA.*

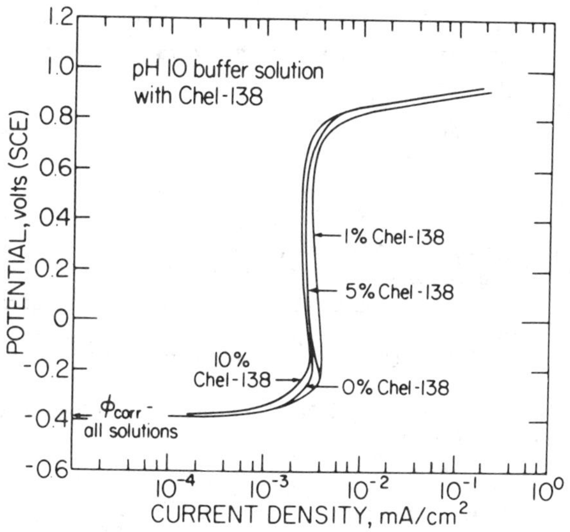

FIG. 4—*Anodic polarization curve for steel in pH 10 buffer solution as a function of volume percent Chel-138.*

concentration. This was, however, the only feature of these agents which was adverse with regard to corrosion. Because this critical potential was in most instances positive (noble) to the corrosion potential of steel in concrete, it is anticipated that any acceleration of corrosion rate due to either a HEDTA or Chel-138 admixture of the concentrations investigated should be minimal.

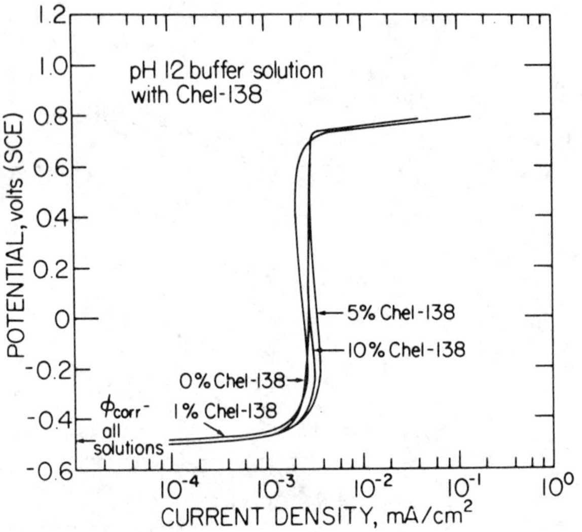

FIG. 5—*Anodic polarization curve for steel in pH 12 buffer solution as a function of volume percent Chel-138.*

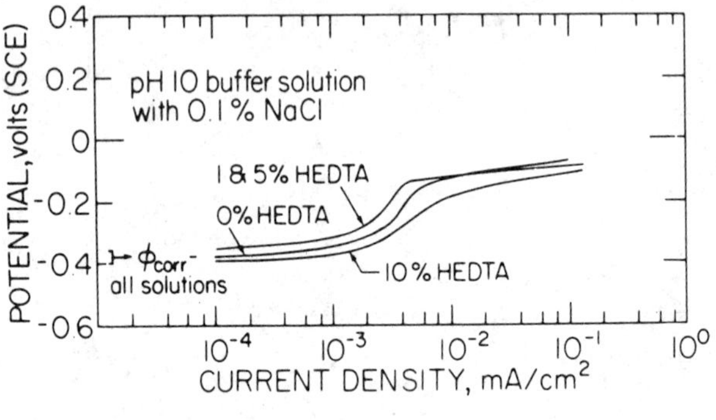

FIG. 6—*Anodic polarization curve for steel in pH 10 buffer solution with 0.1NaCl as a function of volume percent HEDTA.*

Concrete Cracking Experiments

Time-to-cracking data for concrete specimens (mix design and chelating admixture concentrations as in Tables 2 and 3, respectively) are reported in Fig. 10 as a function of the direct anodic current density which was impressed upon the embedded metal. A possible complicating aspect of this data is that No. 5 reinforcing steel was employed in the no-admixture specimens, whereas No. 4 was used in the admixtured ones. A limited number of tests upon No. 4 steel specimens without admixture were performed, however, and no variations as a function of steel diameter were apparent.

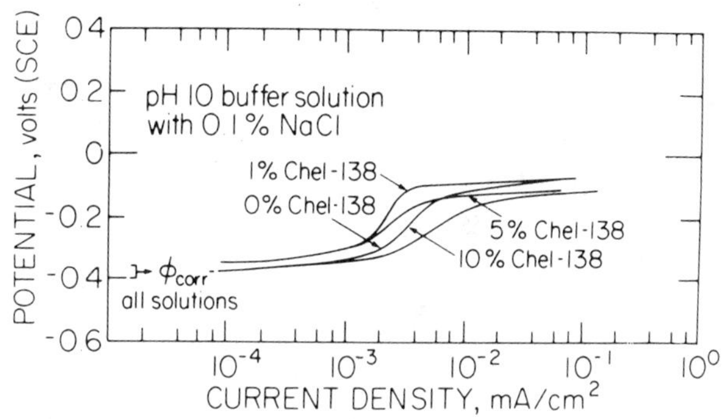

FIG. 7—*Anodic polarization curve for steel in pH 10 buffer solution with 0.1NaCl as a function of volume percent Chel-138.*

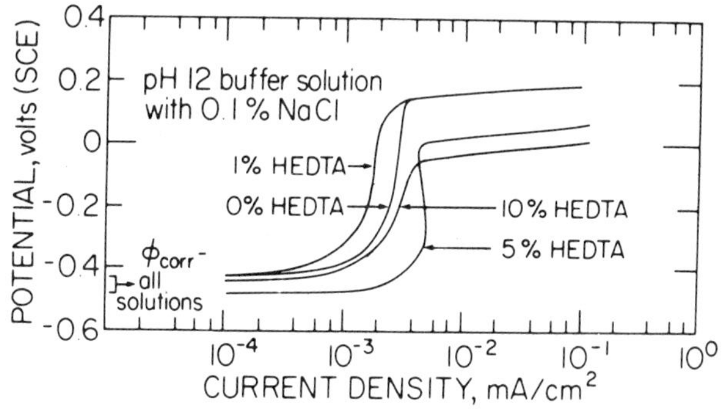

FIG. 8—*Anodic polarization curve for steel in pH 12 buffer solution with 0.1NaCl as a function of volume percent HEDTA.*

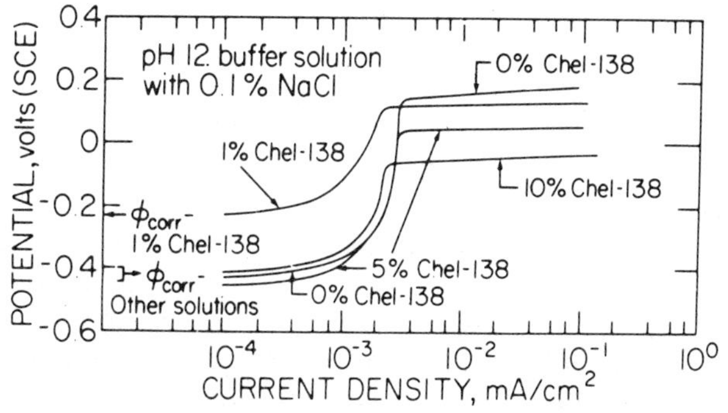

FIG. 9—*Anodic polarization curve for steel in pH 12 buffer solution with 0.1NaCl as a function of volume percent Chel-138.*

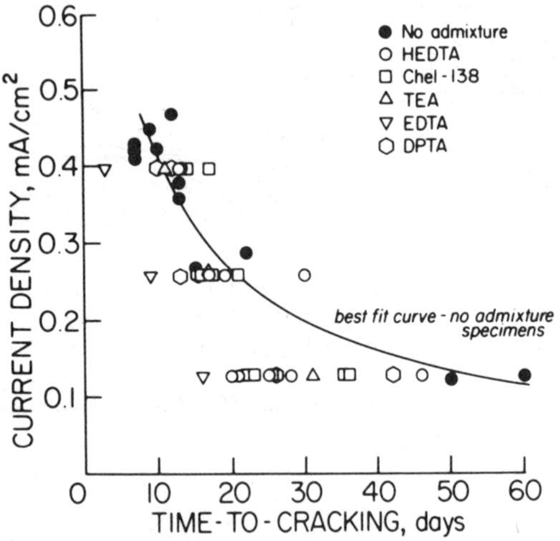

FIG. 10—*Time-to-cracking of reinforced concrete specimens as a function of impressed current density and type of chelating admixture.*

Of particular significance is time-to-cracking for admixtured specimens in comparison with the no-admixture case. While at the highest current density (0.4 mA/cm^2) some of the chelating additions, HEDTA and Chel-138 in particular, have enhanced cracking times, at the lowest current density the no-admixture specimens exhibited the greatest life. This result is disappointing in regard to the usefulness of chelating agents in retarding or inhibiting concrete cracking and spalling.

One explanation for the reduced cracking time of specimens with admixture in comparison to ones without is in terms of concrete strength. Thus, it may be reasoned that the higher the concrete strength, the more resistant the material should be to cracking from tensile hoop stresses resulting from corrosion product accumulation. This is supported by the fact that HEDTA and Chel-138 exhibited the highest compressive strength (see Table 4) and also the greatest cracking time of any type of admixtured specimen. However, ordering of cracking time for specimens with other chelating admixtures does not correlate simply with concrete strength. This suggests that factors besides this latter parameter (concrete strength) are also important.

An additional factor which may influence time-to-cracking of admixtured concrete specimens with respect to nonadmixtured ones is occurrence during testing of an aqueous puddle on the top surface near the metal-concrete interface. The nature of this has been described previously by Grimes et al [*6*]. Interestingly, such puddle accumulation was observed only for specimens without chelating additions. This suggests either that there were fundamental differences in the electrochemical reactions and perhaps in subsequent chemical reactions which evolved the ultimate corrosion product for admixture versus no-admixture specimens, or that there were differences in concrete structure and properties which influenced occurrence of the puddle, or both. One possibility is that reaction of iron ions with the chelating agent formed a corrosion product which remained soluble until it diffused to a higher pH region some distance from the metal-concrete interface.

The presence of this liquid provided a path for the impressed current other than the intended one, which was directly from metal into concrete. If this factor could be taken into account, then the difference in cracking time between the two specimen types might be less or the order of cracking even reversed.

Data for several of the admixture types exhibited scatter, particularly at the lowest current density. A possible explanation for this is in terms of the decreasing slope of the time-to-cracking curve with decreasing current density. As an additional point, it was noted during cracking experiments upon concrete specimens with HEDTA and Chel-138 that specimen life decreased in successive tests. Since all specimens were fabricated at the same time, this suggests that the concrete specimens were less resistant to cracking as time progressed. The effect was most pronounced at the lowest current density investigated (0.13 mA/cm^2). Consequently, these data have been replotted in Fig. 11, where time of the test has been considered. It may be reasoned from this that if the experiments had commenced sooner after pouring the concrete, time-to-cracking for admixtured specimens at the lowest current density may have exceeded that of the nonadmixtured ones. Presumably, the

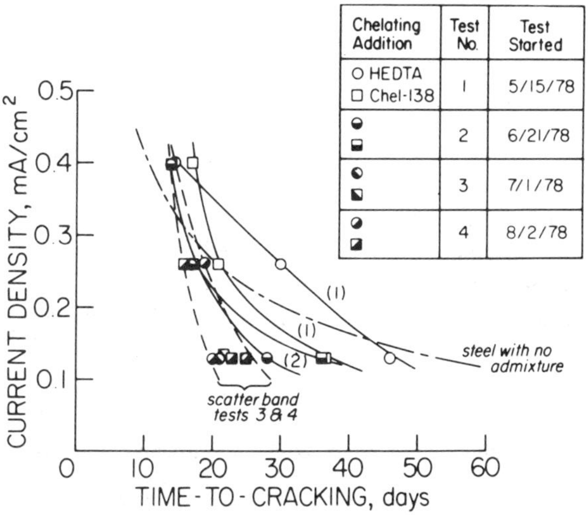

FIG. 11—*Time-to-cracking for the data in Fig. 10, taking into account time the test commenced.*

reduced cracking resistance with increased time is a consequence of some chemical reaction or reactions between the concrete and chelating agent. Additional tests to investigate this phenomenon further have not yet been performed. On this basis it is not unrealistic that cracking time of reinforced concrete members could be prolonged by an appropriate chelating admixture, but this has not been observed because of reactions between these chemicals and the concrete.

Conclusions

1. Each of the chelating agents, TEA, EDTA, DTPA, HEDTA, and Chel-138, enhanced iron ion solubility, the effect being most pronounced for near-neutral pH.
2. All of the chelating admixtures investigated reduced 28-day concrete compressive strength.
3. For the concentrations investigated, neither HEDTA nor Chel-138 significantly influenced the anodic polarization behavior of steel in chloride-free solutions of pH 10 and 12.
4. In 0.1NaCl solutions at pH 10 and 12, both HEDTA and Chel-138 decreased the critical potential for pitting, the effect being more pronounced at pH 10. This decrease was in proportion to the chelating agent concentration.
5. For most of the test conditions, time-to-cracking was less for reinforced concrete specimens with chelating admixture compared with the no-admixture case. This was probably due to reactions between the chelating admixture and concrete. Cracking time could probably be extended if the chelating agent did not reduce concrete strength.

Acknowledgment

The authors wish to express appreciation to the National Oceanographic and Atmospheric Administration Sea Grant Project No. 04-7-158-44046 for financial support. The assistance of Mr. R. P. Brown and of the Florida Department of Transportation for continued interest and support for the overall corrosion research program at Florida Atlantic University and for preparation of specimens used in these tests is acknowledged.

References

[1] Rosa, E. B., McCollum, B., and Peters, O. S., "Electrolysis of Concrete," U.S. Bureau of Standards Technologic Paper No. 18, 1913.

[2] Tremper, B., Beaton, J. L., and Stratfull, R. F., *Highway Research Board Bulletin* 182, 1958, pp. 18–41.

[3] Larsen, T. J., McDaniel, W. H., Brown, R. P., and Sosa, J. L., "Corrosion Inhibiting Properties of Portland and Portland-Pozzolan Cement Concretes," Status Report, Florida Department of Transportation Office of Materials and Research, Jan. 1975.

[4] Spellman, D. L. and Stratfull, R. F., "Concrete Variables and Corrosion Testing," California Department of Public Works, Division of Highways, Materials and Research Department, Report No. 035116-6, Jan. 1972.

[5] Lewis, D. A. and Copenhagen, W. J., S. *African Industrial Chemist*, Vol. 2, 1957, pp. 207–219.

[6] Grimes, W. D., Hartt, W. H., and Turner, D. H., "Cracking of Concrete in Sea Water Due to Embedded Metal Corrosion," to be published in *Corrosion*.

[7] Berman, H. A., "Effect of Sodium Chloride on the Corrosion of Concrete Reinforcing Steel and on the pH of Calcium Hydroxide Solution," Federal Highway Administration Report FHWA-RD-74-1, Jan. 1974.

J. M. Gaidis,[1] *A. M. Rosenberg,*[1] *and I. Saleh*[1]

Improved Test Methods for Determining Corrosion Inhibition by Calcium Nitrite in Concrete

REFERENCE: Gaidis, J. M., Rosenberg, A. M., and Saleh, I., **"Improved Test Methods for Determining Corrosion Inhibition by Calcium Nitrite in Concrete,"** *Corrosion of Reinforcing Steel in Concrete, ASTM STP 713,* D. E. Tonini and J. M. Gaidis, Eds., American Society for Testing and Materials, 1980, pp. 64–74.

ABSTRACT: The use of open-circuit potential measurements for determining the extent of corrosion suffers from the inability to assign a corrosion rating with confidence to an observed voltage taken at a given location on a given substrate. We have attacked the problem by using a microprocessor to accumulate multiple readings on a given deck. The solid-state electronics reads, converts, and stores the data faster and with more reliability than is possible with completely manual operation.

One problem in field applications is attachment to the rebar mat in order to measure the voltage. We report the usefulness of a two-probe method, which can be used without connecting to the metal framework of the bridge. An area large enough to contain active *and* passive regions must be surveyed.

A new series of reinforced concrete decks has been cast with and without calcium nitrite inhibitor to confirm earlier findings that corrosion can be controlled at an addition rate of 2 percent by weight of cement. Also, the results from the construction of the first highway bridge built with calcium nitrite are reported here.

KEY WORDS: corrosion, concretes, reinforcing steels, admixtures, alkali-aggregate reactions, inhibitors, portland cement, strength of materials

Some time ago, we began studying the use of calcium nitrite as a corrosion inhibitor for use in concrete. Several electrochemical test methods were used to show the benefits of using calcium nitrite as an admixture in the concrete to prevent corrosion. The beneficial effects of calcium nitrite on concrete, as well as the corrosion inhibiting results, were reported elsewhere [*1–4*].[2] One of the tests that we are continuing to study is the Accelerated Bridge Deck Corrosion Test.

[1]Senior research chemist, research manager, and applied mathematician, respectively, W. R. Grace & Co., Columbia, Md. 21044.

[2]The italic numbers in brackets refer to the list of references appended to this paper.

In this test, decks were constructed with and without calcium nitrite according to a design of the Federal Highway Administration [5]. The decks were daily ponded with a 3 percent solution of sodium chloride and open-circuit potentials were measured. When potential readings were first taken on the decks, the results appeared inconsistent because only six data points were taken on each surface. In order to understand the effects of the corrosion inhibitor, we eventually decided to take 297 readings on each deck. The 46 decks now in test take considerable time to record and display the values from each deck.

In order to speed up data collection, a microprocessor to record the open-circuit potentials and a computer to display our results in an easily understood fashion have been added to our data collection process.

In the first series of decks that were made, we found that in the concrete with higher compressive strength the protection calcium nitrite offered the steel was more pronounced. Therefore, we have constructed a new series of decks to clearly demonstrate the effects of calcium nitrite in quality concrete. In one part of the new series a retarder was used to closely parallel conditions of use. In another part, a superplasticizer was used to substantially reduce the water used in the concrete. A statistically designed experiment was utilized in order to minimize effects of nonuniformity in the concrete.

Although these deck tests represent the best laboratory approximation of field conditions, there are still substantial differences between these and actual bridge decks in a natural environment. In these accelerated tests, the steel is two inches on centers and a salt solution is put on the decks each day of the year. In this instance, corrosion is at its worst in the summertime because of the high temperature and salt concentration. However, we have made an actual bridge deck with calcium nitrite to obtain field results.

We have also adapted our method of measuring open-circuit potentials in order to measure the corrosion on a bridge deck without requiring excavation to attach the voltmeter to the reinforcing steel below the concrete.

The Accelerated Bridge Deck Corrosion Test with the Corrosion Inhibitor and Other Admixtures

In 1977 we made concrete decks according to the Accelerated Bridge Deck Corrosion Test developed primarily by the Federal Highway Administration [5]. The decks [1.8 m by 0.6 m by 15.2 cm (72 by 24 by 6 in.)] were salted daily with a 3 percent sodium chloride solution. A section of one of the decks containing no nitrite was examined for determining a relationship between open-circuit potential and chloride ion content at the reinforcing bar. Figure 1 shows that as the chloride penetration increases, the potential increases.

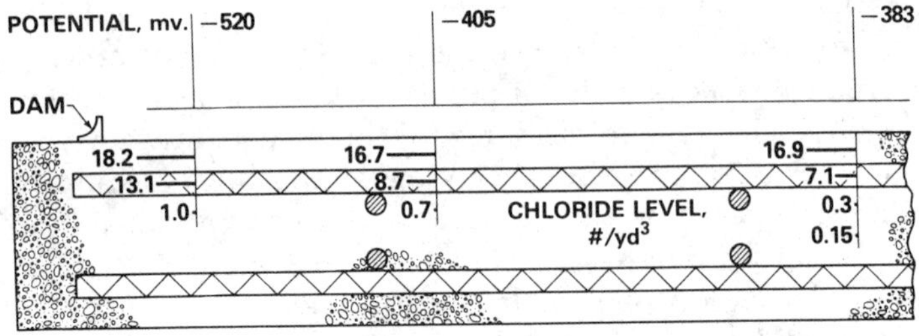

FIG. 1—*Chloride levels and potentials for Slab No. 8.*

A 2^2 full factorial design was selected to study the effects of two variables on corrosion. The two variables were cement content and amount of calcium nitrite. Each run was replicated three times with a different truckload of mix for each deck. Table 1 lists the 12-run design in order (randomized to minimize systematic errors), with results available to date.

In this series, an attempt was made to control setting time changes due to the corrosion inhibitor. Previously, it was found that calcium nitrite was an accelerator of set but that this acceleration could be offset by the use of a retarder.

Another series of decks was made using a superplasticizer to reduce the amount of water used in the mix as well as the corrosion inhibitor. It is believed that the reduced permeability of the concrete due to the water reduction plus the corrosion inhibitor will produce a superior concrete structure. Table 2 gives the results of the concreting.

The trucks were brought to the site "dry" (that is, containing only aggregate moisture) and admixtures were added halfway through the water addition: Daravair[3] air-entraining agent (based on vinsol resin), then WRDA-19[3] superplasticizer, followed by Darex Corrosion Inhibitor (DCI) and the rest of the water. In these batches, the slump [ASTM Test for Slump of Portland Cement Concrete (C 143-74)] was kept low in order to maintain the water/cement ratios given in Table 2. The air content was maintained at 6 ± 1 percent. The cement content was 418 kg/m³ (7½ sacks/yd³). All of the concrete was ready-mix. Each condition represents one truckload. One slab was made from each batch for the Accelerated Bridge Deck Corrosion Test. It is interesting to note that on previous days other batches were made with a retarder only and the same cement factor. The average compressive strength at 28 days was 30.2 MPa (4382 psi). This was 9 percent lower than the lowest in the series with the superplasticizer and 68 percent lower than the highest.

[3]Registered trademark.

TABLE 1—*Experimental design: initial data on concrete decks.*

Design Run Order	Cement Content, sacks/yd^3		Calcium Nitrite % by Weight of Cement	Slump, in.	Air, %	Compressive Strength, psi		
	Design	Actual				3-day	7-day	28-day
1	6.5	6.5	0	3	7.0	2767	3290	4190
2	6.5	6.7	2	2¾	5.2	4020	5027	6207
3	7.5	7.6	0	2½	7.2	2820	3317	4160
4	7.5	7.5	0	3¾	7.6	3203	3620	4590
5	6.5	6.8	2	3	5.2	3593	4927	6210
6	6.5	6.7	2	2¾	5.8	1950	4923	5943
7	7.5	7.8	2	2¼	5.6	4880	6607	8000
8	6.5	6.5	0	2¾	10.5	2343	3083	3820
9	7.5	7.6	0	3¼	8.5	3147	3997	5000
10	7.5	7.8	2	2¼	5.6	4880	6607	8000
11	7.5	7.8	2	2¼	5.9	3773	5877	7133
12	6.5	6.5	0	2¾	10.5	2343	3083	3820

NOTE: 1 yd^3 = 0.76 m^3; 1 in. = 2.54 cm; 1 psi = 6.895 kPa.

TABLE 2—*Concrete decks made with superplasticizer and corrosion inhibitor.*

Admixtures (% by Weight of Cement)				Compressive Strength, psi[c]		
DCI[a]	WRDA-19[b]	Water/Cement	Air, %	3 days	7 days	28 days
...	0.38	0.34	6.1	3563	4120	4787
...	0.57	0.305	6.8	4250	4823	5897
...	0.76	0.28	7.9	5033	5670	6707
2.0	0.38	0.31	5.0	4593	5440	6657
2.0	0.57	0.34	5.0	4870	5857	7097
2.0	0.76	0.30	6.6	5347	6367	7370

[a] DCI or Darex Corrosion Inhibitor is a 30 percent solution of calcium nitrite. The weight of the solids in the solution is considered here.
[b] WRDA-19 is a sodium naphthalene sulfonate formaldehyde resin (Type B superplasticizer).
[c] 1 psi = 6.895 kPa.

A Rapid Technique for Measuring Open-Circuit Potentials on Concrete Decks

In our earlier corrosion measurements, we found that a clear interpretation could be obtained only if we measured the potential across the entire surface. A 9 × 33 grid was placed on the deck while potentials were being measured. In this way, 297 values were obtained for each deck.

In our earlier experiments a digital voltmeter in conjunction with a copper-copper sulfate half-cell was used to obtain corrosion potential measurements on each deck. These data were then punched on International Business Machines (IBM) cards and loaded into a large computer for deck mapping and other analyses. Manual measurements, reading of the digital voltmeter, and recording of the data on a 9 × 33 mapped sheet representing the deck required a time period of 30 to 40 min per deck. The keypunching and verification of the data on IBM cards required an additional 25 to 30 min per deck.

A microprocessor has now replaced the manual readout and recording of our data. Use of the microprocessor has reduced the measuring time by a factor of 5 or 6 from the required time of 30 or 40 min to 5 to 8 min per deck. It has also eliminated the keypunching of the IBM cards and, therefore, has saved the keypunching and verification time. In addition, it has eliminated the human error in data recording and keypunching.

The microprocessor system consists of the following components (schematized in Fig. 2):

1. *KIM-1 microprocessor board*—including a keyboard and input/output I/O terminal ports.
2. *Cassette tape recorder*—for loading program instructions into the microprocessor board.

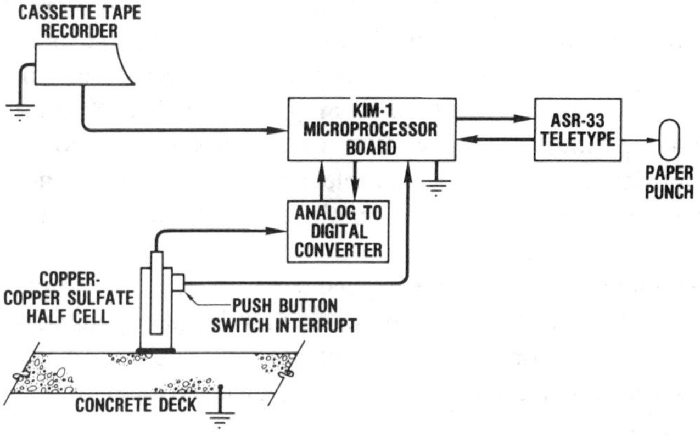

FIG. 2—*Schematic of data recording with microprocessor.*

3. *Analog-to-digital converter*—for converting the analog corrosion potential into its digital equivalent.
4. *Copper-copper sulfate half-cell*—for measuring corrosion potential.
5. *Push button switch attached to the copper-copper sulfate half-cell*—for sending interrupt signals to the microprocessor.
6. *Buzzer*—to alert the operator for different tasks.
7. *ASR-33 teletype with paper punch capability*—for punching microprocessor data onto paper tape. The teletype is also used for communication with the processor as an alternate to the keyboard provided on the KIM-1 board.

All components, except the cell and teletypewriter, are placed inside a portable 43 by 28 by 10-cm (17 by 11 by 4 in.) attaché case.

Program instructions are loaded from the cassette recorder into the microprocessor board. Then, the copper-copper sulfate half-cell is placed on the concrete deck and the pushbutton switch is depressed. This provides a signal to the microprocessor to collect the reading from the particular location on the deck. The processor in turn signals the analog-to-digital converter to convert the analog potential signal, supplied by the cell, to its equivalent digital value. This value is then stored in the memory location of the microprocessor preassigned to a corresponding location of the deck. Data are measured in 9 rows of 33 points each. At the end of each 33 points, the buzzer sounds once to signal the end of a row, and at the end of the 9th row it sounds twice to signal the end of the deck measurements.

At the end of the measurements, the deck number is punched on the paper tape and the processor is instructed to dump its data onto the paper tape. This punched paper tape replaces the cards, which were manually punched, and is later read into the main computer.

Data in the main computer are analyzed for deck uniformities and average corrosion levels. Different decks are compared and a number of plots indicating deck ranges and variations are generated along with histograms and other useful plots. A map of each deck is plotted in its original numeric values, along with a map indicating the hot (corroded) spots on the deck. It is then a simple matter to color in the different areas on the computer printout to obtain an easily understood picture of the corrosion on each deck.

Since the major cost of these experiments, after deck preparation, is the manpower needed for measurements and keypunching of the data, using the microprocessor has enabled us to generate and measure more decks to identify other important variables such as water-to-cement ratio and cement content.

The New Hampshire Test Bridge

In the laboratory testing we have run thus far, field use has been simulated in order to obtain results in a reasonable length of time. There really is no substitute for actual use of the product in the field to determine its efficacy. On 26 Sept. 1978, concrete containing calcium nitrite was placed in a bridge deck on Route 122, north of Hollis, N.H. It is a two-lane structure with an asphalt wearing surface and is approximately 9 m (30 ft) long. This bridge was built with exposed concrete curbs on both lanes. The depth of the concrete is 38 cm (15 in.) with two reinforcing mats made with No. 8 steel bars. (See Table 3.)

The corrosion inhibitor was put in the eastbound substructure concrete at an addition rate of 2 percent by weight of cement. The westbound lane did not contain this additive.

Monitoring of corrosion will be conducted by the New Hampshire Department of Transportation. Their tests will include visual inspection of the wearing surface and curbs. They will also determine chloride ion content and steel corrosion tendencies. One way to follow the corrosion history of this bridge without introducing damage due to testing is described in the next section.

Double-Probe Corrosion Potential Method

During a 12-month period, it is planned to use calcium nitrite in 10 or more bridges throughout the United States. The way the bridge will be constructed and monitored will, of course, be the responsibility of the agency constructing the structure. However, we felt that a nondestructive method of measuring the open-circuit potential of these bridges would be helpful in studying the benefits of calcium nitrite in these bridges.

TABLE 3—*Concrete data from the New Hampshire test bridge.*

Mix Design			
Cement		752 lb	
Sand		1200 lb	
¾-in. gravel		1660 lb	
Water added		27 gal	
Darex AEA		1 fl oz/cwt cement	
Daratard-HC		1 fl oz/cwt cement	
DCI		15 lb	
Wet Concrete Properties			
Slump		3.0 in.	
Air content		5.5%	
Unit weight		144.8 lb/ft^3	
Yield		27.1 ft^3/yd^3	
Compressive Strength, psi			
1 day	7 days	14 days	28 days
1167	4059	5093	5535

NOTE: 1 lb = 0.45 kg; 1 gal = 3.8 litres; 1 fl oz = 30 ml; 1 ft^3 = 0.03 m^3; 1 yd^3 = 0.76 m^3.

From the Tafel equation, we know that the change in measured potential is proportional to a function of the corrosion current [*3*]. Thus, if we could measure the change in potential from one area of the deck to another, the relative corrosion rate could be determined. It is assumed that the change in potential from area to area can be related to the change in potential that has occurred where a highly active potential is found. It is assumed that shortly after construction of the bridge all areas are passive.

In the usual method of measuring open-circuit potentials, one side of a high-resistance voltmeter is directly attached to the reinforcing steel and the other is attached to a copper-copper sulfate half-cell. The copper-copper sulfate electrode is then moved across the concrete as the potential is measured at each location.

In the double-probe technique, two copper-copper sulfate electrodes are attached to the voltmeter, and then both electrodes are placed on the concrete surface so that the differences in potential from area to area can be measured.

It is true that the absolute potential can be measured only by attaching the voltmeter directly to the reinforcing; however, corrosion starts slowly and generally begins in a small area. Thus, to measure the onset of corrosion on new bridges, we feel this technique will be helpful in pinpointing the troublesome areas. Eventually, as corrosion progresses and becomes more uniform, this technique will not be able to discern the differences. However, by the time the readings have progressed that far, other signs of corrosion would be apparent, such as cracking and rust staining.

To show the type of results that could be obtained with the double-probe technique, we have taken a deck showing only a slight corrosion tendency and measured the open-circuit potential over the entire surface by both methods.

Figure 3 shows the potential grid obtained by the conventional method where the voltmeter is attached to the reinforcing steel directly.

The double-probe method uses one electrode as a reference, moving the other over the surface to measure the difference in potential. Figure 4 shows a map obtained with the reference electrode placed on a low potential area. Figure 5 shows the map obtained with the reference electrode placed on a high potential area. All three maps show the potential spread from the

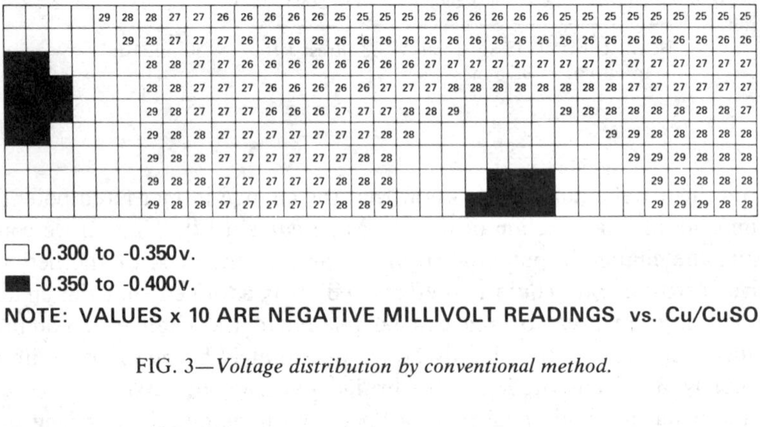

□ -0.300 to -0.350v.

■ -0.350 to -0.400v.

NOTE: VALUES x 10 ARE NEGATIVE MILLIVOLT READINGS vs. $Cu/CuSO_4$

FIG. 3—*Voltage distribution by conventional method.*

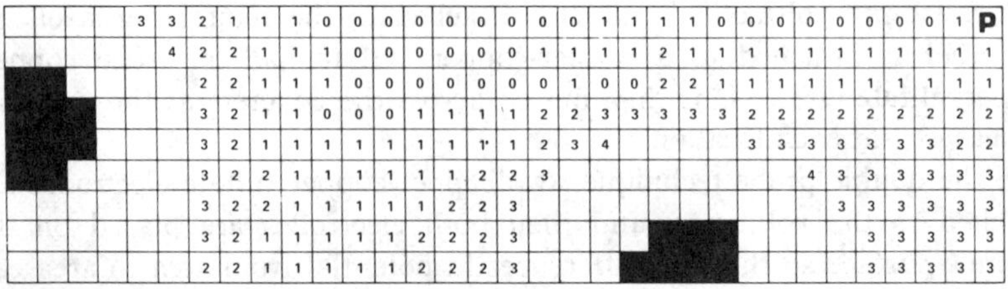

P is Position of Reference Electrode

■ **Exceeds Reference Voltage by -0.100v.**

□ **Exceeds Reference Voltage by -0.050v.**

NOTE: VALUES x 10 ARE NEGATIVE MILLIVOLT READINGS vs. $Cu/CuSO_4$

FIG. 4—*Voltage distribution by two-probe method (little or no corrosion at reference electrode position).*

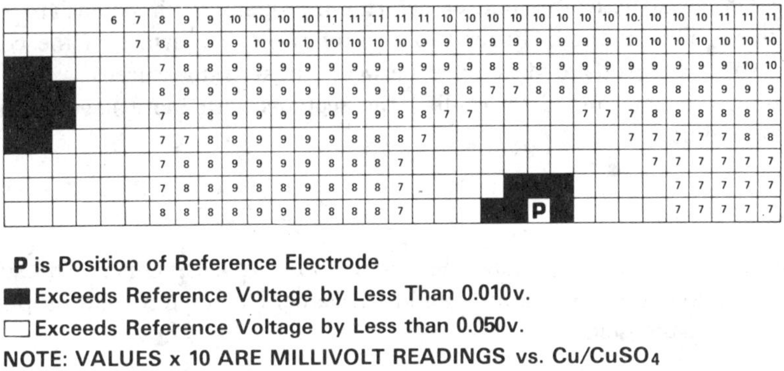

FIG. 5—*Voltage distribution by two-probe method (corrosion at reference electrode position).*

passive areas to the areas showing signs of corrosion to be slightly greater than 0.1 V. The double-probe method can determine only differences in potential, but this still allows the most negative (most active) areas to be located, even in situations where connection to the underlying steel is not possible at all.

It might occur that certain parts of the reinforcing in a bridge are not connected. If this happens, there will be a sudden jump in potential, leading to spurious readings. By placing the reference electrode and the probe close to each other in the area where a suspected erroneous reading was obtained, one can determine the validity of the values. If the readings were correct, the voltmeter should read zero or close to that; if the steel is disconnected, the voltmeter will continue to read erratically.

Conclusions

In our studies on the effects of calcium nitrite in concrete, we have used the Accelerated Bridge Deck Corrosion Test. Previously, we found that the beneficial effects of calcium nitrite were more pronounced when the concrete compressive strength was higher. A new series of decks has been designed and built to test this. The strength of concrete using a superplasticizer and the corrosion inhibitor was found to be exceptionally high.

Because of the large number of potential readings needed on the decks we have thus far constructed, we now use a microprocessor to read and record our data. A computer prints out the data in an easily understood chart.

A field test was undertaken to determine the benefits of calcium nitrite under actual conditions.

In order to rapidly determine the condition of this field test and others, a rapid nondestructive method of determining open-circuit potentials was developed. This method utilizes two electrodes on the surface of the concrete where the voltage difference between them is measured. Areas of corrosion are determined by finding the sections with more negative potential readings.

Acknowledgments

The authors wish to acknowledge the help of Dr. José Giner, who suggested the double-probe method, and Mr. Daniel Harrington, who supplied the details of the field test.

References

[1] Rosenberg, A. M., Gaidis, J. M., Kossivas, T. G., and Previte, R. W. in *Chloride Corrosion of Steel in Concrete, ASTM STP 629*, American Society for Testing and Materials, 1977, pp. 89–99.
[2] Lundquist, J. T., Jr., Rosenberg, A. M., and Gaidis, J. M., *Materials Performance*, Vol. 18, March 1979, pp. 36–40.
[3] Rosenberg, A. M. and Gaidis, J. M., *Materials Performance*, Vol. 18, Nov. 1979, pp. 45–48.
[4] Rosenberg, A. M. and Gaidis, J. M., Transportation Research Record No. 692, 1978, pp. 28–34.
[5] Clear, K. C. and Hay, R. E., Federal Highway Administration Report #FHWA-RD-73-32, Washington, D.C., April 1973.

R. Rider[1] *and R. Heidersbach*[2]

Degradation of Metal-Fiber-Reinforced Concrete Exposed to a Marine Environment

REFERENCE: Rider, R. and Heidersbach, R., **"Degradation of Metal-Fiber-Reinforced Concrete Exposed to a Marine Environment,"** *Corrosion of Reinforcing Steel in Concrete, ASTM STP 713,* D. E. Tonini and J. M. Gaidis, Eds., American Society for Testing and Materials, 1980, pp. 75–92.

ABSTRACT: The object of this research was to determine the effects of a marine environment on the integrity of metal-fiber-reinforced concrete. Metal-fiber-reinforced concrete has potential for uses in marine structures where the metal fiber may introduce tensile strength, abrasion resistance, and fatigue properties which may justify the added cost when compared with conventional concrete. Metal-fiber-reinforced concrete specimens were tested in flowing seawater and freshwater laboratory exposures. Comparisons were made to specimens exposed in the tidal zone of Narragansett Bay. Freeze-thaw experiments were also conducted. Results were obtained using standard and modified ASTM testing procedures as well as electrochemical corrosion rate monitoring techniques. The results indicate that stainless steel fibers are needed in marine applications.

KEY WORDS: metal-fiber-reinforced concrete, corrosion, steel, seawater, reinforcing steel, cement, concrete, marine

Concrete is the most widely used man-made construction material in the world. Low cost, versatility, and adequate compressive strengths are reasons for the popularity of concrete construction. The principal disadvantages of concrete are low tensile strength and brittleness.

Metal-fiber-reinforced concrete can improve these tensile and brittle properties. Randomly oriented short metal fibers have been shown to increase tensile strength and retard crack propagation. Table 1 lists the improvements which can be obtained in concrete by the addition of metal fibers. A typical fibrous concrete mix design incorporates approximately

[1] Associate professor, Department of Ocean Engineering, University of Rhode Island, Kingston R.I. 02881.
[2] Civil engineer, Brown and Root, Co., Houston, Tex. 77001.

TABLE 1—*Attainable properties of 2 volume percent fibrous concrete* [5].

Property	Increase Over Plain Concrete, %
First-crack flexural strength	50
Ultimate modulus of rupture strength	100
Ultimate compressive strength	15
Ultimate shear strength	75
Flexural fatigue endurance limit	225
Impact resistance	325
Sandblast abrasion resistance index	200
Heat spalling resistance index	300
Freeze-thaw durability index	200
General Advantages of Fibrous Concrete	
Much greater resistance to cracking	
Improved fatigue characteristics	
Far superior resistance to thermal shock	
Significantly thinner sections for a given design	
Elimination or reduction of other types of reinforcing materials	
Increased production rate with thinner sections	
Less maintenance and longer life	
Isotropic properties	

1 to 1½ volume percent metal fibers. Fibers 0.04 cm (0.016 in.) in diameter and 2.5 cm (1 in.) long are typical.

Metal-fiber-reinforced concrete requires higher cement contents and smaller aggregate size than conventional concrete. Quality control is more important because of difficulties in achieving random fiber orientation. Fibrous concrete has been successfully used in airport runways and bridge overlays [*1–6*].[3]

Marine applications for concrete have increased in recent years. Concrete is generally inert in seawater and has low maintenance requirements. Problems can appear in the tidal and splash zones where chemical attack of the cement paste, corrosion of reinforcing steel, cyclic freezing and thawing, and wear and abrasion are concentrated [*4–9*]. Petroleum drilling and storage platforms, ships, and barges are currently being constructed from concrete.

This study was initiated to test the suitability of fibrous concrete for marine applications. The mechanical strength and abrasion advantages listed in Table 1 indicate that fibrous concrete might offer significant advantages for certain marine applications. Before these advantages can be utilized, the questions of corrosion susceptibility and freeze-thaw durability must be resolved. A series of controlled corrosion experiments and freeze-thaw exposures was conducted in seawater in an attempt to answer these questions.

[3] The italic numbers in brackets refer to the list of references appended to this paper.

Marine Concrete Degradation

A number of chemical reactions must be controlled to produce durable concrete for marine applications.

Sulfates present in seawater are especially damaging to concrete. They react with tricalcium aluminate (C_3A) in the cement paste to form calcium sulfoaluminate hydrate (ettringite) [*10–14*]. Ettringite has more than twice the volume of the tricalcium aluminate, and this volume expansion can cause swelling and cracking.

Another reaction that occurs in sulfate-containing waters is the formation of gypsum (calcium sulfate) from calcium hydroxide present in the cement paste [*9,17*]. Gypsum is soft and can be removed by wave action or currents. This removal exposes fresh concrete to attack.

The chlorides in seawater increase the solubility of these reaction products. This can lower the expansion caused by these reactions [*8*] and make them less damaging in seawater than they are in freshwater or groundwaters. Nonetheless, chemical attack is a significant problem for marine concrete. Limitations on tricalcium aluminate, high cement contents, nonreactive aggregates, and low water/cement ratios are all used in marine concretes to minimize the effects of sulfate attack.

Reinforcing steel in concrete can corrode in the presence of moisture and oxygen. Chloride intrusion accelerates this attack [*7,15–21*], probably due to a lowering of the electrical resistivity of moist concrete, which serves as the electrolyte in any galvanic cell which may form. Figure 1 illustrates the reactions which can occur during the corrosion of reinforcing steel.

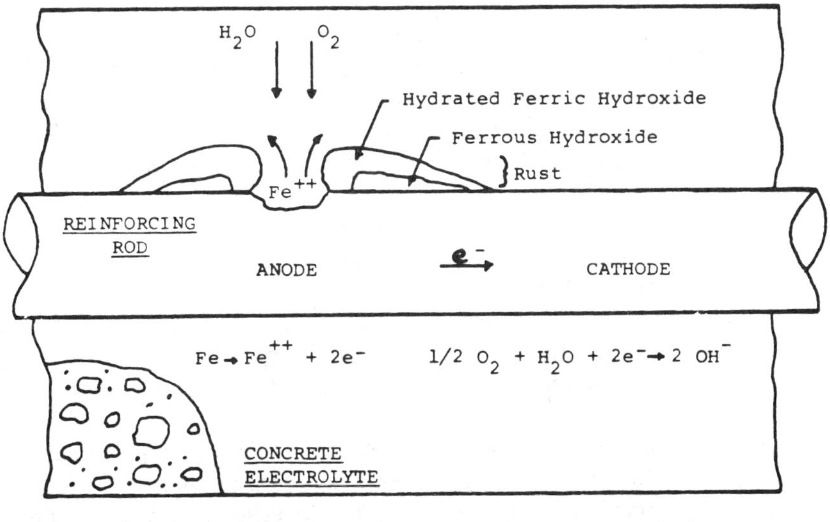

FIG. 1—*Galvanic corrosion of concrete reinforcing steel.* [19]

The iron reaction products that form due to corrosion have a greater volume than the steel from which they were formed. This can lead to cracking, which, in turn, promotes moisture ingress and further corrosion.

Corrosion of steels in concrete is minimized by lowering the water/cement ratio, which lowers moisture permeability, and by specifying a minimum concrete cover depth to provide a moisture barrier for the steel [*8,16,19,22*].

Freeze-thaw deterioration causes cracking in wet concrete subjected to repeated freezing cycles [*1,9,13,23–27*]. Air entrainment is the accepted method for counteracting this phenomenon.

It is possible that fibrous concrete may arrest cracking and counteract freeze-thaw deterioration. This possible benefit must be balanced against the increased moisture and sulfate ingress which might be caused by the large number of metal-concrete interfaces, which could accelerate water and dissolved ion migration [*14,27*]. The metal fibers reach to the concrete surface and thus have no protective cover. Thus it would appear that fibrous concrete has possible advantages in a marine environment that experiences freezing and thawing. It may also have corrosion disadvantages which could prevent its successful use in saltwater applications.

One problem which must be noted is that fibrous concrete produces stiff concrete mixes. This precludes the use of low water/cement ratios. It has been suggested that this difficulty may be overcome by the use of water-reducing admixtures [*28*]. No reports of investigations into this possibility have appeared [*29*].

Experimental Program and Results

The foregoing discussion indicates that marine exposures may cause degradation of metal-fiber-reinforced concrete due to sulfate attack, metal corrosion, and cyclic freezing and thawing. Previous studies of corrosion pointed out the need for careful exposure control [*11,30*], and it was decided to use natural seawater in the chemical/corrosion exposures in order to determine any synergistic effects caused by fouling and suspended marine organisms.

The necessity for using relatively high water/cement ratios in fibrous concrete has already been mentioned. Other alterations from conventional concrete mix design include:

1. Limitation of maximum aggregate size to 0.95 cm (3/8 in.) in order to reduce friction between the aggregate and thc fibers.
2. Use of high cement contents in order to fully coat the fiber surfaces with cement paste.
3. Special fiber addition and concrete consolidation techniques to insure random orientation and prevent fiber agglomeration, or balling.

Table 2 presents the mix designs used in this investigation. A fiber content of 1.25 volume percent was used, although higher percents are common

TABLE 2—*Concrete mix proportions.*

Mix[a]	Cement/ Type[b]	Water	Sand[c]	Coarse Aggregate[d]	Fiber[e]	W/C[f]	Slump[g]	Air Content, %
300 H	680 IA	367	1576	1052	159 S	0.54	2	4.5
400 H	680 IA	347	1576	1052	162 M[h]	0.51	1	4.5
500 H	680 IA	320	1576	1052	0	0.47	3/4	4.5
600 H	680 II	347	1576	1052	171 SS	0.51	1/4	3.3
700 H	680 IA	347	1576	1052	159 S	0.51	5/8	4.0
900 M	680 IA	340	1483	1178	0	0.50	5 3/8	7.1
1000 M	680 IA	408	1483	1178	159 S	0.60	8	9.5
1100 M	680 IA	340	1483	1178	159 S	0.50	3	6.5

[a] H = hand mixed; M = machine mixed.
[b] All constituents in lb/yd^3.
[c] Fineness modulus = 2.80.
[d] 3/8 maximum aggregate.
[e] M = Meltex (0.016 in. dia., 1 in. long (nominal)) 1.25% by volume.
S = Low carbon steel (0.016 in. dia., 1 in. long) 1.25% by volume.
SS = 302 stainless steel (0.013 in. dia., 1 in. long) 1.25% by volume.
[f] W/C = water/cement.
[g] Slump in in.
[h] Meltex is a trade name of the National Standard Company and refers to an 18-8 stainless steel fiber.

in practice [*4–5*]. The objective of this study was to determine degradation after relatively short test exposures. Low fiber content, low cement contents, and the use of Type I cement, instead of the Types II or V normally specified for marine use, were deliberately incorporated into this investigation to maximize any degradation tendencies that might occur.

Three types of metal fibers were used in the investigation. Chopped carbon steel fibers with a brass drawing lubricant are typical of the fibers commonly used in fibrous concrete construction. Stainless steel fibers were also used to test their possible increased resistance to corrosion. An experimental melt-spun stainless steel fiber was also tested. It should be noted that the volume percent of fibers in all mixes was held constant. The reduced diameter of the stainless steel fibers necessitated a larger number of fibers per unit of volume. This resulted in different cement/metal interface areas per unit of volume. Similar considerations hold for the Meltex[4] fibers. Thus the mechanical properties of concrete made with each type of fiber were substantially different. Alterations in mechanical properties, and possibly moisture ingress characteristics and degradation, cannot be compared between the specimens made with the different types of fibers used in this study.

Flowing Seawater Exposures

Exposure beams measuring 3.75 by 3.75 by 35 cm (1.5 by 1.5 by 14 in.) were cut from 7.5 by 7.5 by 70-cm (3 by 3 by 28 in.) cast beams. The larger beams were required for casting in order to insure random fiber orientation. The smaller beams were used as exposure specimens in order to maximize the area-to-volume ratio. This would emphasize any deterioration effects which might occur during the limited exposure times available for this study [*15,31*].

The beams were exposed to air-saturated flowing seawater at ambient temperature. Figure 2 illustrates the exposure arrangement. The beams were broken below the waterline, above the waterline, and at the waterline. This procedure yielded three sets of data for each exposure specimen and was intended to clarify the differences occurring in the three locations.

Figure 3 shows the results of flexure tests for exposure of up to 12 months. A trend of degradation may be apparent, but the results indicated are well within the experimental reproducibility of concrete. Thus these results cannot be interpreted to mean that the specimens degraded measurably during the exposure period investigated. Other specimens tested in flexure and compression also failed to show any significant trends. Similar results have been obtained in studies conducted in other laboratories [*11,30*]. Nonetheless, Fig. 4 shows that corrosion did occur on exposure beams containing carbon steel fibers. Thus it must be concluded that the limited

[4] Registered trade name, the National Standard Co., Niles, Mich.

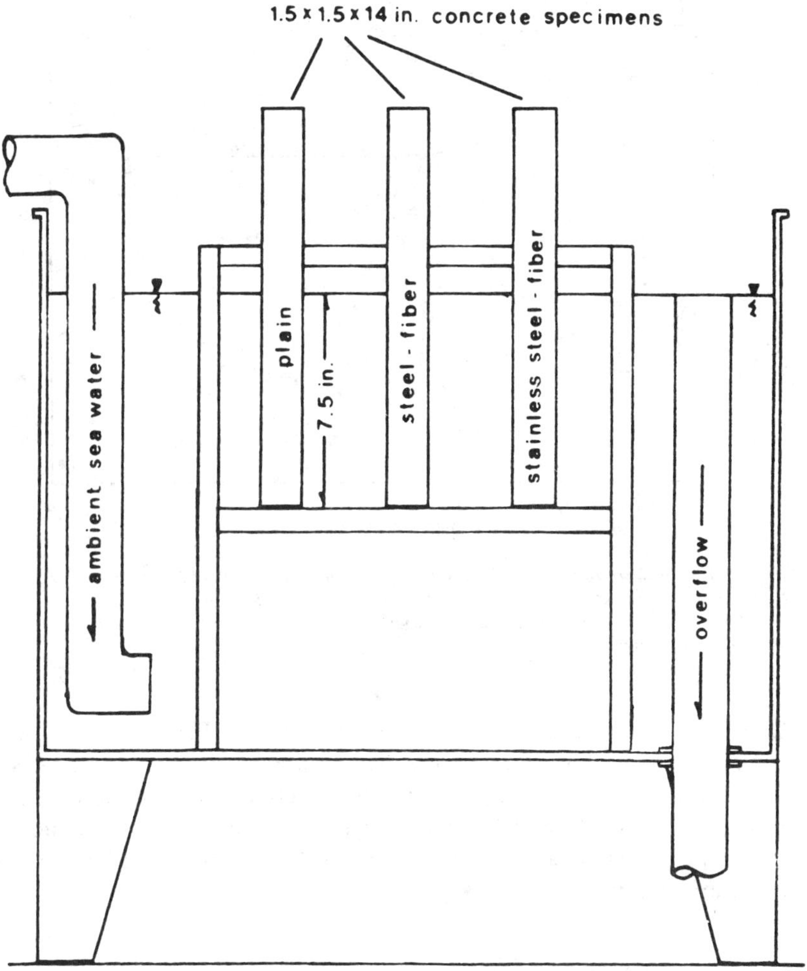

FIG. 2—*Flow-through seawater exposure arrangement (1 in. = 2.54 cm).*

exposure times (one year and less) which have been reported here and elsewhere have been insufficient to produce measurable strength changes using the relatively insensitive mechanical properties tests commonly used for concrete testing. Obviously other methods are needed to measure this effect.

Figure 5 is a scanning electron micrograph of a carbon steel fiber removed from a position approximately 1.27 cm (½ in.) from the surface of a seawater exposure specimen. The smooth rounded mineral product in the center was identified by X-ray spectroscopy to be an iron corrosion product. The surface of a stainless steel fiber removed from a similar location on a

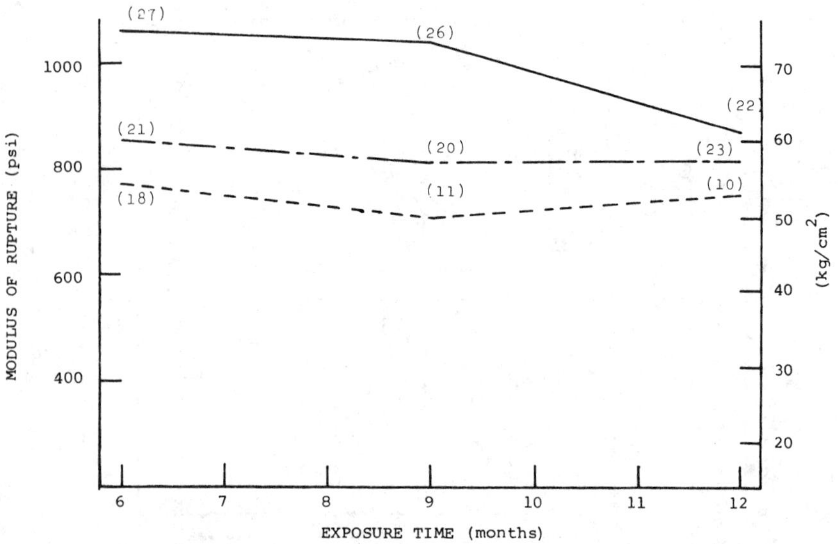

FIG. 3—*Flexural strength of carbon steel fiber-reinforced beams [3.8 by 3.8 by 35.6 cm (1½ by 1½ by 14 in.)] exposed to flowing seawater. ——break above waterline; —-— break at waterline; --- break below waterline. Numbers in ( ) are the number of fibers crossing the break below the neutral axis.*

different specimen exposed to seawater is shown in Fig. 6. The mineral shown on the surface was tentatively identified as ettringite by comparison with scanning electron micrographs published elsewhere [*32*]. All exposure specimens were investigated for evidence of corrosion. No evidence of corrosion was found on any of the stainless steel fibers examined by scanning electron microscopy.

Freeze-Thaw Experiments

Four hundred cycles of freezing and thawing were conducted on a series of air-entrained fiber-reinforced concrete specimens. Reference *39* describes these tests, which were modified from standard ASTM tests, in detail. No measurable degradation was noted on the nonreinforced control specimens or on the fiber specimens, so no conclusions about the effect of fibers can be made.

Linear Polarization Measurements

An attempt was made to measure the corrosion behavior of a continuous wire reinforced concrete specimen using the exposure apparatus and specimen arrangement shown in Fig. 7. Instrumentation shown in Fig. 8 was

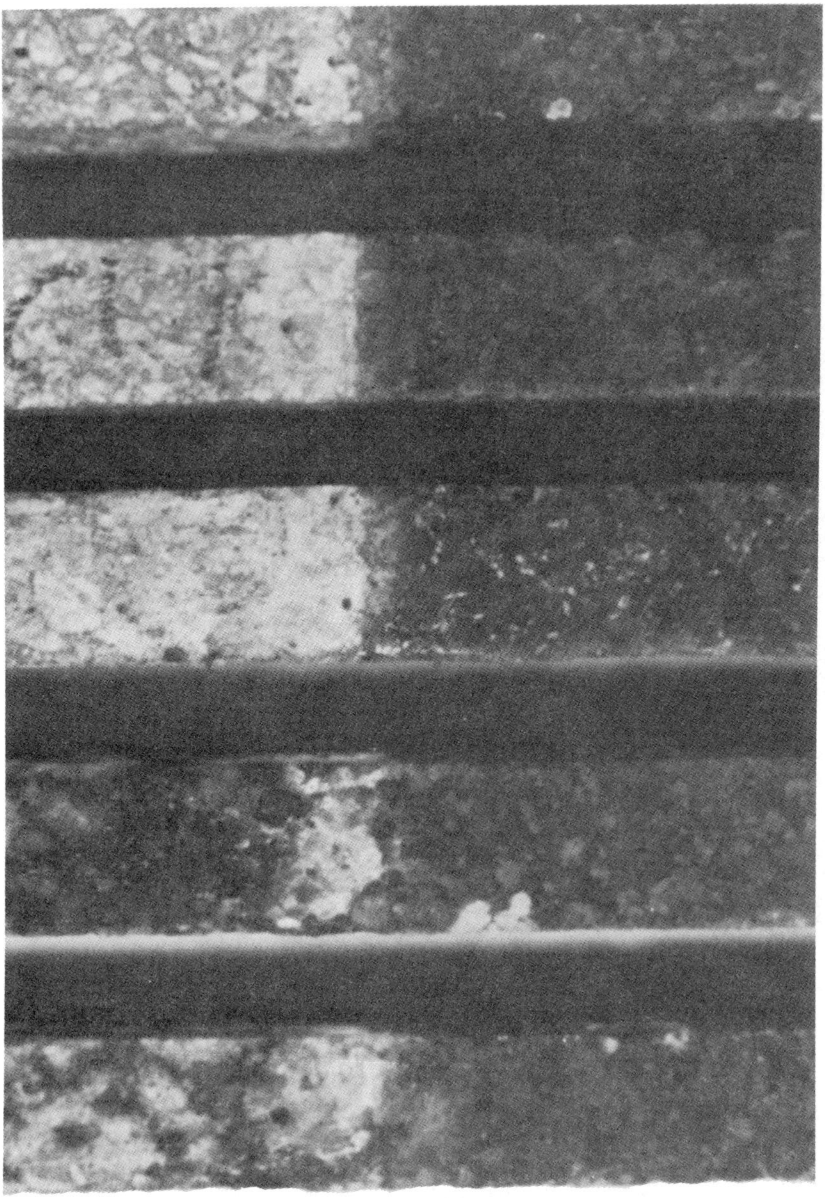

FIG. 4—*Twelve-month flow-through seawater exposure specimens [3.8 by 3.8 by 35.6 cm (1½ by 1½ by 14 in.)]. Cut side up. From the top: stainless steel, plain, Meltex, carbon steel, and carbon steel fiber reinforced.*

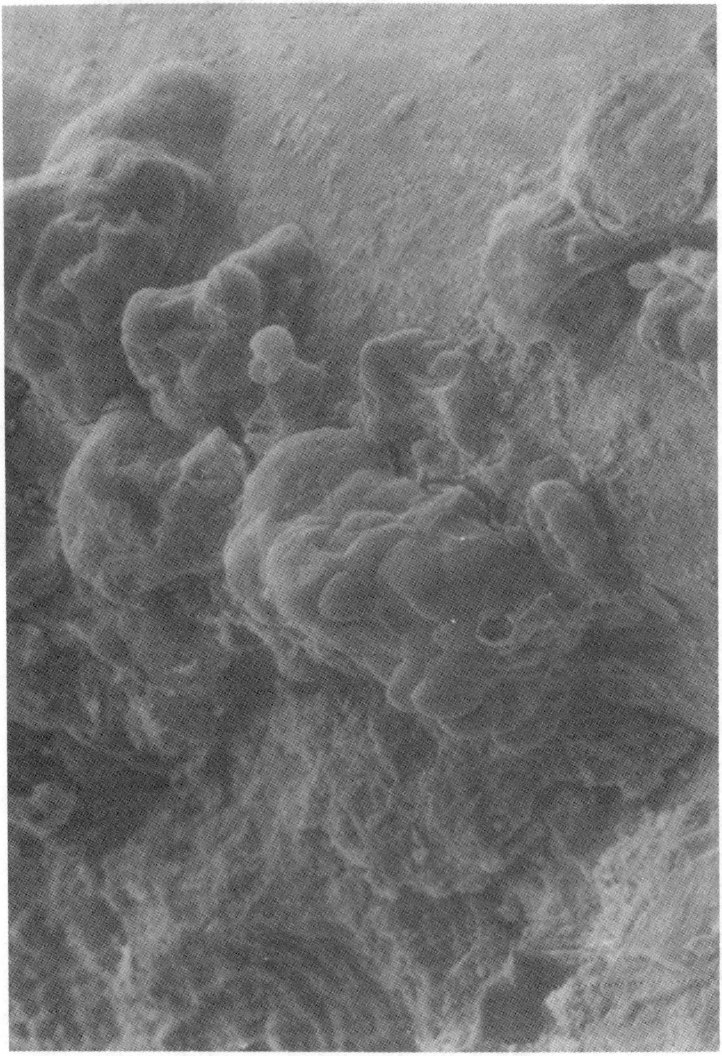

FIG. 5—*Low-carbon-steel fiber from the surface of a concrete specimen exposed to seawater for six months. Smooth area to the right is the original wire; rounded shapes in the center are rust; grainy mass in the upper left contains calcium with some silicon; angular surface in lower left contains silicon with some aluminum and calcium* (×425).

FIG. 6—*Cement/metal interface on a 302 stainless steel fiber from the surface of a concrete specimen exposed to seawater for six months* ($\times$2550). *Lower right area is base metal; rest of surface is covered with cement; needles are probably ettringite.*

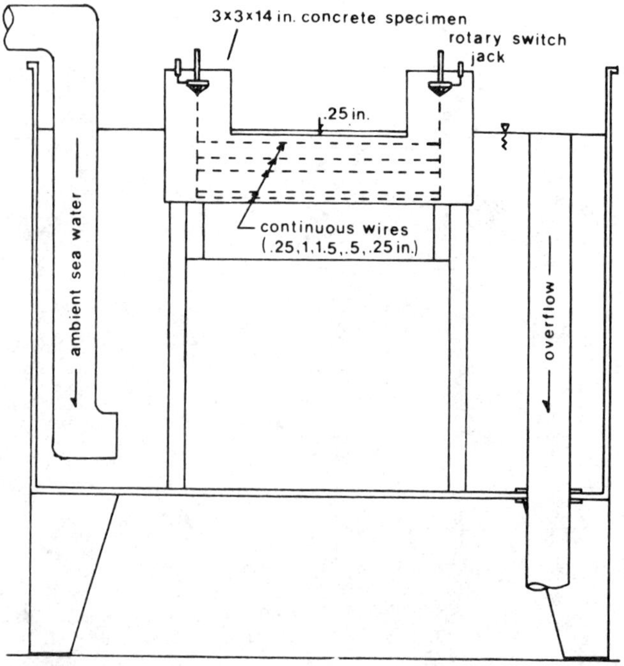

FIG. 7—*Exposure arrangement for continuous wires used to measure corrosion rates electrically.*

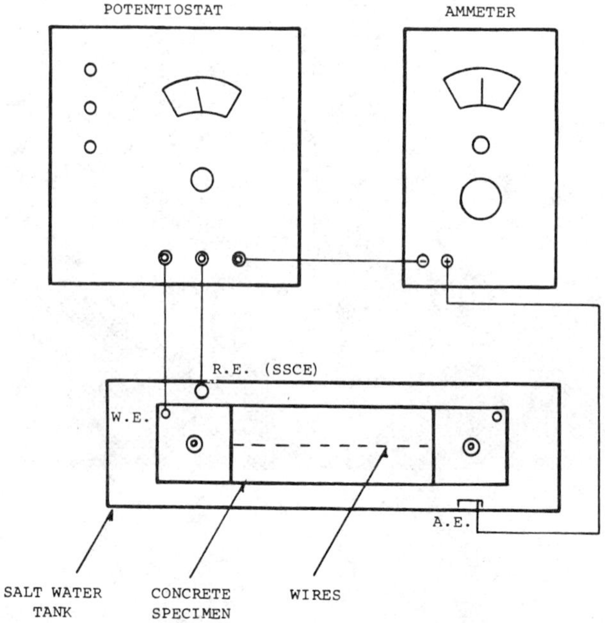

FIG. 8—*Arrangement for making linear polarization measurements.*

used to measure instantaneous corrosion rates using the linear polarization techniques introduced by Stern and Geary [*33–38*]. Detailed results are reported elsewhere [*39*], but Fig. 9 shows typical results. The corrosion rate can be roughly correlated with depth of cover. Steep slopes are indicative of low corrosion rates. Only two of the eight specimens represented in Fig. 9 showed appreciable corrosion rates. Table 3 summarizes the corrosion rate results obtained in this research using linear polarization techniques.

The Stern-Geary technique can be used to calculate an approximate corrosion rate. These rates were calculated for comparison purposes and are reported elsewhere [*39*]. Corrosion rates are not shown here because they are based on the assumption that corrosion rates are uniform over the exposed surface area. Steel in concrete tends to corrode by pitting (see Fig. 5), and thus average depth of attack calculations would be mis-

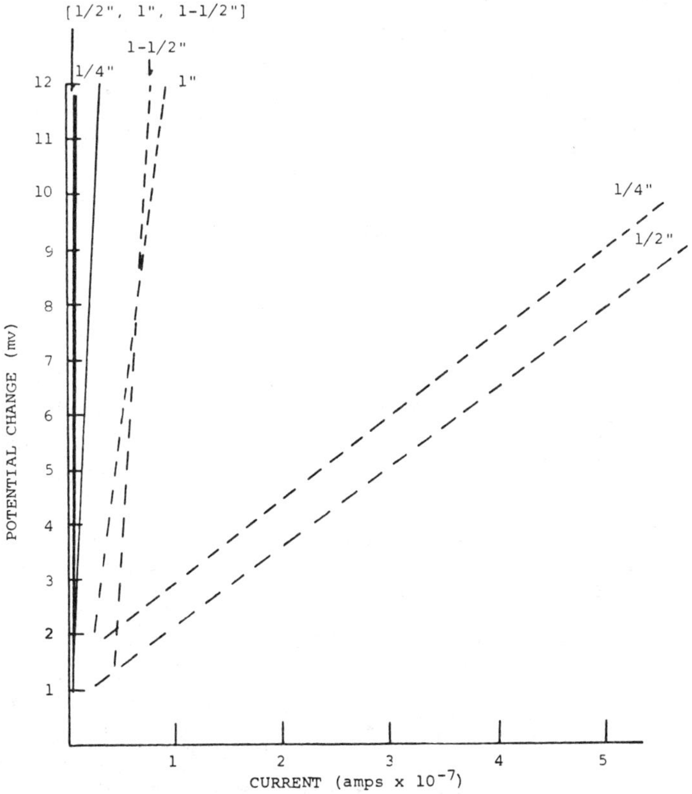

FIG. 9—*Linear polarization data for carbon steel wires embedded in carbon steel fiber-reinforced concrete (water/cement ratio = 0.6). Specimens and wires identified by depth of concrete cover. Specimens --- exposed 196 days in unpainted concrete. Specimens — exposed 210 days in damp-proofed (painted) concrete.*

TABLE 3—*Calculated corrosion rates for 20.3 cm (8 in.) of carbon steel wire [0.038 cm ϕ (0.015 in. ϕ)] in painted and unpainted carbon steel fiber-reinforced concrete exposed to flowing seawater for 6½ months.*

Specimen	Wire Depth,[b] in.	Days Immersed	Rest Potential[a]		Current Density, μa/cm²	Corrosion Rate	
			Before mV	After, mV		mpy[c]	mdd
Unpainted							
1011 W1	¼ top			wire broken			
W5	¼ bottom	196	422	423	0.21	0.302	1.652
W4	½ bottom	196	300	308	0.17	0.317	1.734
W2	1 top	196	197	203	0.019	0.031	0.167
W3	1½ middle	196	152	149	0.011	0.014	0.078
Painted							
1013 W1	¼ top	210	218	228		no slope	
W5	¼ bottom	210	161	171	0.011	0.011	0.057
W4	½ bottom	210	173	178	0.004	0.004	0.023
W2	1 top	210	168	172	0.002	0.003	0.014
W3	1½ middle	210	167	176	0.004	0.004	0.020

[a] Ag/AgCl electrode.
[b] 1 in. = 2.54 cm.
[c] mpy = mils/year.

FIG. 10—*7.6 by 7.6 by 106.7 cm (3 by 3 by 42 in.) stainless steel figer-reinforced beam after 12 months' exposure in Narragansett Bay. Left face exposed to waves.*

leading. Nonetheless, the linear polarization techniques seems to offer promise for indicating corrosion of metal embedded in concrete where other techniques are too insensitive to detect corrosion activity.

Tidal Zone Exposures

A series of larger [7.6 by 7.6 by 106.7 cm (3 by 3 by 42 in.)] concrete beams was exposed in the tidal zone under a pier at the University of Rhode Island's Narragansett Bay Campus for a period of 12 months. The reinforced control specimen was lost due to winter storms. The carbon steel specimen was visibly corroded.

Figure 10 shows the stainless steel specimen. The crack in the center of the specimen was caused by a storm shortly after placement, and salt water penetrated the entire width of the specimen. Nonetheless the fibers did not corrode. After one year of exposure the beam was broken in flexure. The calculated flexure strength of 4433 kPa (643 psi) was approximately the same as an unreinforced beam would have before exposure.

TABLE 4—*Suggested mix design for metal-fiber-reinforced concrete to be used in ocean structures.*

	Quantity	
Constituent	kg/m^3	lb/yd^3
Cement—Type II	519	875
Water	234	394
Fine aggregate	761	1282
Coarse aggregate—max ⅜ in.	607	1024
Fibers—2 volume %, carbon steel/stainless steel 0.016 ϕ × 1 in.	148	250
Admixtures		
air-entraining agent		
water reducing agent		
Mix Properties		
slump	7.6 to 12.7 cm	3 to 5 in.
air content		6 to 7.5%
Precautions		
avoid segregation, excessive bleeding, and over-vibration.		
cure sufficiently		
Other Possible Considerations		
surface coatings (water proofers/damp proofers)		
polymer inpregnation		
interval sealants		

Conclusions

Metal-fiber-reinforced concrete can be used in a marine environment. While carbon steel fibers are used for on-shore construction, stainless steel should be used in marine structures. Table 4 represents a suggested mix design for marine applications of fibrous concrete.

Acknowledgments

This project was funded by the University of Rhode Island Sea Grant Program. The National-Standard Company supplied the metal fibers. Flowing seawater exposure tanks were made available at the Environmental Protection Agency Narragansett Laboratory.

References

[*1*] Waddell, J. J., *Concrete Construction Handbook,* 2nd ed., McGraw-Hill, New York, 1974.

[*2*] *Design and Control of Concrete Mixtures,* 11th ed., Portland Cement Association, Ill., 1968.

[*3*] Gerwick, B. C., in *Proceedings,* American Society Civil Engineers, March 1971, pp. 1–16.

[*4*] Gray, B. H., Williamson, G. R., and Batson, G. B., "Fibrous Concrete Construction Material for the Seventies," Construction Engineering Research Laboratory, Ill., May 1972.

[*5*] Williamson, G. R. and Gray, B. H., "Technical Information Pamphlet on Use of Fibrous Concrete," Construction Engineering Research Laboratory, Ill., May 1973.

[*6*] Hanna, A. N., "Steel Fiber Reinforced Concrete Properties and Resurfacing Applications," Portland Cement Association, Ill., 1977.

[*7*] *Proceedings,* Symposium on Concrete Construction in Aqueous Environments, ACI Publication SP-8, American Concrete Institute, Detroit, Mich., 1962.

[*8*] Biczok, I., *Concrete Corrosion and Concrete Protection,* Chemical Publishing Co., Inc., New York, 1967.

[*9*] Mather, B., "Effects of Sea Water on Concrete," U.S. Army Engineer Waterways Experiment Station, Vicksburg, Miss., Miscellaneous Paper No. 6-690, Dec. 1964.

[*10*] Mather, B., "Field and Laboratory Studies of the Sulfate Resistance of Concrete," U.S. Army Waterways Experiment Station, Vicksburg, Miss., Miscellaneous Paper 6-922, Aug. 1967.

[*11*] Shroff, J. K., "The Effect of a Corrosive Environment on Properties of Steel Fiber Reinforced Portland Cement Mortar," M.S. Thesis, Clarkson College, Potsdam, N.Y., Sept. 1966.

[*12*] "Building Code Requirements for Reinforced Concrete," ACI Standard 318-63, American Concrete Institute, Detroit, Mich., June 1963.

[*13*] Mehta, P. K. and Williamson, R. B., "Durability of Cement Concrete in Sulfate Environment," University of California, Berkeley, Calif., AD 767 312, Sept. 1973.

[*14*] Lankard, D. R., Corrosion/76, National Association of Corrosion Engineers, Houston, Tex., Paper No. 16, March 1976.

[*15*] *1973 Book of ASTM Standards,* "Concrete and Mineral Aggregates," Part 10, American Society for Testing and Materials, Sept. 1973.

[*16*] Robinson, R. C., *Materials Protection and Performance,* Vol. 11, No. 3, Ameron South Gate, Calif., March 1972.

[*17*] *Concrete Manual,* U.S. Department of the Interior, Bureau of Reclamation, 7th ed., 1966.

[*18*] Gouda, V. K. and Mourad, H. M., *Corrosion Science,* Vol. 15, No. 5, May 1975.

[19] Mozer, J., Biachini, A., and Kesler, C., "Corrosion of Steel Reinforcement in Concrete," T&AM Report No. 259, Department of Theoretical and Applied Mechanics, University of Illinois, Urbana, Ill., April 1964.
[20] Houston, J., Atimtay, E., and Ferguson, P., "Corrosion of Reinforcing Steel Embedded in Structural Concrete," Research Report 112-1F, Center for Highway Research, University of Texas, Austin, Tex., March 1972.
[21] Halvorsen, G., Kesler, D., Robinson, A., and Stout, J., "Durability and Physical Properties of Steel Fiber Reinforced Concrete," Report No. DOT-TST 76T-21, U.S. Department of Transportation, Aug. 1976.
[22] Woods, H., "Durability of Concrete Construction," American Concrete Institute Monograph No. 4, 1968.
[23] Zolin, B., American Institute of Chemical Engineers, New York, Nov. 1977.
[24] Mather, B., "Field and Laboratory Studies of the Sulfate Resistance of Concrete," U.S. Army Engineer Waterway Experiment Station, Vicksburg, Miss., Miscellaneous Paper No. 6-922, Aug. 1967.
[25] Powers, T. C., "Freezing Effects in Concrete," SP 47-1, American Concrete Institute, Detroit, Mich., 1975.
[26] Lyse, I., *Journal of the American Concrete Institute,* June 1961.
[27] Gjorv, O. E., "Durability of Marine Concrete Structures Under Arctic Conditions," Port and Ocean Engineering Under Arctic Conditions Symposium, Tblisi, 1976.
[28] Whiting, D., Portland Concrete Association, Skokie, Ill., private communication, March 1976.
[29] Rose, T., Rider, R., and Heidersbach, R., "Corrosion of Metals in Concrete," University of Rhode Island Marine Technical Report 58, Kingston, R.I., 1977.
[30] Batson, G. B., "Strength of Steel Fiber Concrete in Adverse Environments," U.S. Army Construction Engineering Research Laboratory, Champaign, Ill., Special Report M-218, June 1977.
[31] Cook, H. K. in *Proceedings,* Third Conference on Coastal Engineering, Tblisi, Oct. 1952.
[32] Schittgrund, G. D., "Bond Strength Between Cement Hydrates and Steel," Ph.D. Thesis, University of Illinois, Urbana, Ill., 1975.
[33] *Handbook on Corrosion Testing and Evaluation,* W. H. Ailor, Ed., Wiley, New York, 1971.
[34] Pye, E., "Electrochemical/Instrumental Methods of Determining Corrosion Rates of Buried Metal Structures," Report C-16, American Concrete Institute, Feb. 1977.
[35] Pye, E., "Practical Considerations in Performing Linear Polarization Measurements," Report C-10, American Concrete Institute, 1977.
[36] Hausler, R. H. in *Corrosion,* National Association of Corrosion Engineers, Vol. 33, No. 4, April 1977.
[37] Schaschl, E., "Modern Electrical Methods for Determining Corrosion Rates," National Association of Corrosion Engineers Publication 3D170, 1976.
[38] Fontana, M. and Greene, N., *Corrosion Engineering,* Wiley, New York, 1977.
[39] Rider, R., M.S. Thesis, University of Rhode Island, Kingston, R.I., 1978.

L. D. Flick[1] *and J. P. Lloyd*[2]

Corrosion of Steel in Internally Sealed Concrete Beams Under Load

REFERENCE: Flick, L. D. and Lloyd, J. P., **"Corrosion of Steel in Internally Sealed Concrete Beams Under Load,"** *Corrosion of Reinforcing Steel in Concrete, ASTM STP 713* D. E. Tonini and J. M. Gaidis, Eds., American Society for Testing and Materials, 1980, pp. 93-101.

ABSTRACT: The ability of internally sealed concrete to prevent corrosion in reinforcement under service conditions has not yet been established. To simulate service conditions, members containing flexural cracks under either static or repeated load and subjected to a period of flushing with fresh water were considered. Corrosion of the reinforcement was induced by daily applications of a 3 percent sodium chloride solution.

The results based on half-cell potentials showed that the corrosion of the reinforcement in internally sealed concrete exposed to salt water was dependent on the type of loading. Under static load, no corrosion of the steel occurred. Under repeated load, however, the corrosion of the steel in internally sealed beams and in beams of conventional concrete occurred at the same time. This indicates that internally sealed concrete may offer only limited protection for members subjected to repeated loads. The lack of correlation between half-cell potentials and rate of corrosion suggests that further research in this area is appropriate.

KEY WORDS: internally sealed concrete, corrosion, sodium chloride, repeated loading, static loading, flexural cracks

The internal sealing of reinforced concrete by the addition of wax beads to the mix is a method that has been developed to greatly reduce the permeability of concrete [*1,2*].[3] In addition to low permeability, the material properties of internally sealed concrete are reported to be similar to or better than those of a conventional 6 percent air-entrained concrete [*1*].

The technique of internally sealing concrete with wax beads has been standardized by the Federal Highway Administration (FHWA) [*2*], and only a brief review of the process is given here. The wax beads, which are

[1] Research engineer, Amoco Production Co., Tulsa, Okla.

[2] Associate professor, School of Civil Engineering, Oklahoma State University, Stillwater, Okla. 74078.

[3] The italic numbers in brackets refer to the list of references appended to this paper.

the size of fine sand, are added on a solid volume basis by replacing approximately 7.8 percent of the volume of concrete occupied by fine aggregate. The wax is mixed with the other components of the concrete until it is well dispersed in the mix; the concrete is then placed and cured in a normal manner. After the characteristic strength is achieved, the concrete is heated, which causes the wax to melt and flow into capillaries and pores. Five to nine hours of heating are required for a bridge deck if melting to a 76-mm (3 in.) depth is desired.

Previous studies on the permeability of this concrete were conducted on uncracked unloaded concrete slabs that were ponded daily with 3 percent sodium chloride (NaCl). In one study [*1*], 417 days of NaCl applied did not result in chloride penetration beyond a depth of 22 mm (3/4 in.). In another study [*3*], half-cell readings were obtained using copper-copper sulfate electrodes (CSE) to determine whether reinforcement was corroding. In all cases the potentials remained more positive than −0.20 V CSE during a 30-day period, which indicated that the reinforcement was passive.

In this study the corrosion resistance provided by internally sealed concrete was investigated under simulated service conditions. In particular, the test program investigated beams with flexural cracks subjected to either static or repeated loading. In addition, some of the beams were subjected to a daily period of freshwater flushing before the initiation of corrosion tests.

Because previous studies had demonstrated the impermeable nature of internally sealed concrete, this study involved beams with flexural cracks to determine whether corrosion could be initiated by the intrusion of chloride into the concrete at the cracks. For much the same reason, repeated loading was introduced into the program to determine whether the pulsating width of cracks under such loading would accelerate the penetration of chloride to the level of the steel. A period of freshwater flushing before initiation of corrosion tests was provided to investigate the influence of leaching of the lime at the location of flexura! cracks.

Experimental

Specimens and Materials

The test program involved eight reinforced beams and four plain beams. The reinforced beams were 3.86 m (126 in.) long, 0.203 m (8 in.) wide, and 0.203 m (8 in.) deep. Two No. 4 deformed bars with a yield strength of 338 MPa (49 ksi) were placed in each beam. The bars were spaced at 140-mm (5½ in.) centers and provided with 25 mm (1 in.) of clear cover. The unreinforced beams, which were 1.32 m (52 in.) long, 0.203 m (8 in.) wide, and 0.203 m (8 in.) deep, were used to monitor salt penetration. One half of the beams were of normal non-air-entrained concrete and one-half were of wax bead concrete.

Type I portland cement was used for all concrete. The aggregate consisted of crushed rock with a specific gravity of 2.76, and natural sand with a fineness modulus of 2.33 and a specific gravity of 2.56. The wax beads, which were a blend of 25 percent montan, a naturally occurring ester wax, and 75 percent paraffin, met FHWA guidelines [2] and had a specific gravity of 0.92.

All concrete had a water-cement ratio of 0.5 and a cement factor of 5.4 N of cement per m^3 (7.2 bags of cement per yd^3). Slump ranged from 57 to 90 mm (2¼ to 3½ in.) while air content averaged 2 percent. Table 1 gives concrete strength at 28 days and at the time of test initiation.

Details of Fabrication

All beams were cast in the laboratory using a horizontal nontilting drum mixer with a capacity of 0.45 m^3 (16 ft^3). The beams were cast in well-oiled wooden forms, and control cylinders were cast in vertical cardboard molds. All specimens were cured under plastic film for 24 h before being removed from the forms. Two cylinders from each batch were then placed in a moist room for an additional 27 days. All other specimens were cured under wet burlap for an additional six days and then stored in the laboratory under ambient conditions. The specimens containing wax beads were heat-treated after six weeks of storage.

An insulated plywood enclosure containing infrared heat lamps was used for the heat treatment. Measurements from the thermistors at the surface and interior of the beams were monitored to insure that heating followed the FHWA guidelines [2]. Approximately 6 h were required for the heat treatment. The temperature rise of the control cylinders was not monitored.

The data in Table 1 show that the control cylinders were weaker after the heat treatment than the control specimens tested at 28 days. This suggests

TABLE 1—*Properties of concrete.*

		Compressive Strength, MPa^2 (ksi)			
Batch	Wax	28-Day Moist Cure	At Test Initiation	Age at Test, days	Unit Weight, Mg/m^3 (lb/ft^3)
I	no	41.9 (6.08)	50.7 (7.36)	70	2.45 (153)
II	no	42.6 (6.19)	46.6 (6.76)	69	2.43 (152)
III	yes	36.0[a](5.23)	31.7 (4.60)	68	2.24 (140)
IV	yes	32.4[a](4.70)	28.5 (4.14)	67	2.26 (141)

[a] *Strength without heat treatment.*

that the cylinders, with a higher surface-to-volume ratio than the beams, experienced an excessive rate of temperature rise which resulted in internal cracking.

Testing Procedures

The reinforced beams were supported and loaded symmetrically. Supports of hard rubber were 1.21 m (48 in.) apart and applied loads were 3.04 m (120 in.) apart. While being positioned for test, the beams were rotated 180 deg from the as-cast orientation, which placed the reinforcement in the upper face. Loads of 6.67 kN (1500 lb) were applied either statically or repeatedly to the beams. Static loads were delivered by blocks of concrete, and repeated loads were applied using air cylinders controlled by timers and an air solenoid. The maximum steel stress resulting for the loading was 138 MPa (20 ksi). During periods of cyclic loading, the load was applied in approximately 8 s, held constant for 6 s, and removed in 6 s, which resulted in a period of approximately 20 s.

Immediately after the application of load, crack widths were measured with an optical comparator and found to be 0.08 and 0.13 mm (0.003 to 0.005 in.) in size. The crack widths did not appear to be influenced significantly by the type of concrete. The spacing of cracks averaged 140 and 170 mm (5½ and 6¾ in.) in the case of the beams of internally sealed concrete and normal concrete, respectively.

To allow water to pond on the beam to a depth approximately 3 mm (⅛ in.), a dike of caulking compound was placed on the tensile face; the wetted area was 200 mm (8 in.) wide and 610 mm (24 in.) long, centered between supports.

Both the static and repeated-load regimes involved two beams of internally sealed concrete and two beams of normal concrete. The repeated loading just described was applied for 6 h daily; during the remainder of the day, constant load corresponding to a steel stress of 138 MPa (20 ksi) was present. Before corrosion testing began, a steady flow of tap water was delivered to the tension face of four beams; this condition, hereafter referred to as flushing, was used 6 h each day for 10 weeks. In the case of beams also subjected to repeated loading, the flushing and cyclic loading periods were simultaneous. Corrosion was induced by daily application of 400 ml of a 3.0 NaCl solution. This treatment was performed at the beginning of the daily repeated loading period. The schedule of freshwater flushing and salt water application is given in Table 2.

Instrumentation and Measurements

Half-cell potentials were obtained using a commercially available CSE and a digital voltmeter with an input impedance greater than 1 MΩ. A small piece of moist sponge was placed between the CSE and the surface of the concrete

TABLE 2—*Testing sequence.*

Beam	Type of Concrete	Environment[a] Ten Weeks	Four Weeks	Four Weeks	Six Weeks
1	internally sealed	RL	RL + salt	SL	RL + salt
2	internally sealed	RL + flush	RL + salt	SL	RL + salt
3	normal	RL	RL + salt	SL	RL + salt
4	normal	RL + flush	RL + salt	SL	RL + salt
5	internally sealed	SL	SL + salt	SL	SL + salt
6	internally sealed	SL + flush	SL + salt	SL	SL + salt
7	normal	SL	SL + salt	SL	SL + salt
8	normal	SL + flush	SL + salt	SL	SL + salt

[a] *Explanation of terms:*
RL = 6-h repeated load each day.
SL = 24-h continuous static load.
salt = 3% salt solution each day.
flush = 6-h freshwater flush each day.

to minimize chloride contamination of the electrode. An ion specific probe was utilized to study chloride penetration using procedures developed by the Kansas Department of Transportation [*4*].

Results

Half-Cell Potentials

The maximum negative potentials usually occurred in regions where the NaCl solution was applied. However, for the two sealed beams under static load, voltages remained more positive than −0.20 V CSE, and the most negative voltages were observed at regions near the ends of the beams. Half-cell potentials taken outside the region of salt application were never more negative than −0.25 V CSE; potentials in these locations did not change significantly during the test. Figures 1 and 2 show potential readings versus time for beams under repeated and static loads, respectively. These figures are based on the most negative potential measured in the region of salt application. In these figures −0.35 V CSE is shown as the corrosion threshold; this potential is in agreement with results obtained by Clear [*5*] and Stratfull et al [*6*].

Chloride Penetration

The results of chloride penetration analyses at a depth of 20 mm (¾ in.) are given in Table 3. Since the results obtained with the ion specific probe were not correlated with measurements of chloride content obtained by conventional methods, and the initial quantities of chloride present in the aggregate and mix were not measured, these results are qualitative.

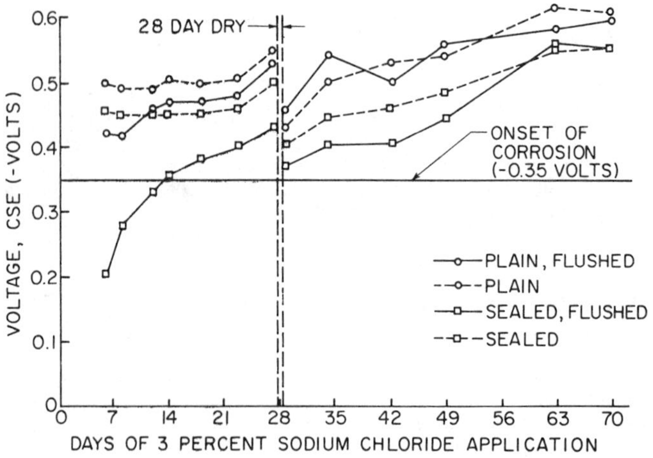

FIG. 1—*Maximum half-cell potentials for the repeatedly loaded beams.*

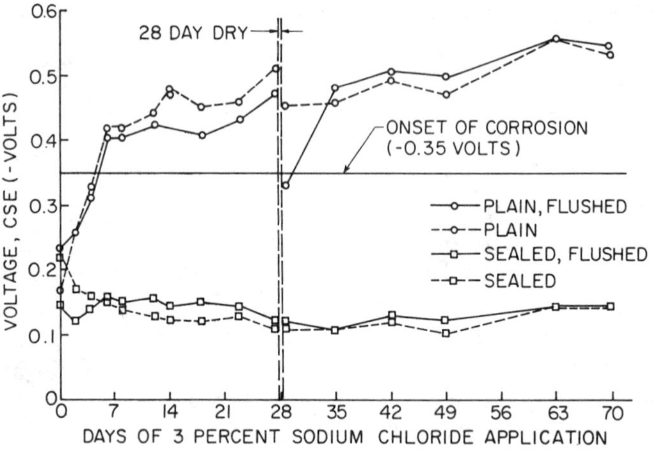

FIG. 2—*Maximum half-cell potentials for the statically loaded beams.*

Leaching of Lime

Specimens Under Static Load—No streaks of lime were developed on the surface of the internally sealed beam; surface tension appeared to block the penetration of water into flexural cracks. The beams of normal concrete developed a small deposit of lime at one crack during the first week of flushing, which did not enlarge during the remainder of the test.

TABLE 3—*Chloride measurements at a 19-mm (3/4 in.) depth.*

No. of Daily Saltings	N of Cl^-/m^3 of Concrete (lb of Cl^-/yd^3 of Concrete)			
	Flushed Beams		Nonflushed Beams	
	Normal Concrete	Internally Sealed	Normal Concrete	Internally Sealed
28	10.5 (1.8)	4.1 (0.7)	14.5 (2.5)	2.3 (0.4)
35	10.5 (1.8)	4.1 (0.7)	13.4 (2.3)	4.1 (0.7)
42	9.9 (1.7)	2.3 (0.4)	13.4 (2.3)	4.1 (0.7)

When the NaCl solution was applied to the beams of normal concrete, it was observed that the solution rapidly seeped through the cracks to the lateral surfaces of the beam which had not been flushed; it appeared that flushing had caused some leached material to be deposited within the flexural cracks, making water entry more difficult.

Specimens Under Repeated Load—Very noticeable quantities of lime were leached from both beams. After two days of flushing, the beam of normal concrete had small ridges of leached material at the edges of all visible cracks; the beam of internally sealed concrete did not reach a similar appearance until about a week later. The deposits on both beams became thicker and wider with time. The material deposited on the sides of the internally sealed beam was more cohesive than that which formed on the surface of the plain beam and also appeared to contain some wax.

When load was removed, the resulting closure of flexural cracks expelled water. Water was expelled from the cracks and ran down the sides of beams made of normal concrete; in many cases the water was milky in appearance. With beams of internally sealed concrete, however, the water—which remained clear—was often drawn back into the cracks when load was reapplied.

Discussion of Results

Influence of Loading

Under static load the internally sealed concrete provided excellent protection against chloride-induced corrosion, as shown in Fig. 2. Under repeated load, however, the pulsating width of flexural cracks acted as a pump drawing the NaCl solution into the beams. Beam 2, which was internally sealed and flushed, did not reach the corrosion threshold as rapidly as the other beams under repeated load; this was possibly a result of a coating of wax or lime formed around the steel during the 10-week period of freshwater flushing.

It should be noted that this study was primarily devoted to the detection of corrosion as evidence by half-cell potentials; these potentials do not necessarily correlate with the rate of corrosion. Therefore, although the beams of normal concrete and internally sealed concrete beams reached the corrosion threshold at about the same time, the severity of corrosion was not established. Another method, such as polarization resistance [*7*], may permit the corrosion rate to be measured in further work.

It is believed that results obtained here indicate that the loading applied to a reinforced concrete structure can significantly influence corrosion. In the future, some studies involving the corrosion of special types of reinforced concrete may obtain results which will accurately relate to in-service performance if members are tested under load.

Leaching of Lime

Significant quantities of calcium carbonate were deposited on the surfaces of the specimens as a result of flushing; larger deposits were obtained under repeated loading. If sufficient lime is leached from the concrete at the location of flexural cracks, carbonation of the remaining lime may cause the internal pH to be lowered, thus rendering reinforcement more susceptible to chloride attack [*8*]. However, no major trend of this type was observed in this study—possibly because of the relatively short duration of the tests.

Conclusions

The experimental results reported in this paper were obtained from an exploratory study of limited scope; however, the following conclusions appear to be justified.

1. The ability of internally sealed concrete to prevent corrosion of embedded reinforcement in the presence of salt water is dependent upon the load environment. Under conditions of static load the reinforcing steel can remain passive; however, under conditions of repeated load where flexural cracks are present, corrosion can occur about as quickly in internally sealed members as in members of normal concrete.

2. The flushing of members with fresh water for 10 weeks before the application of a sodium chloride solution did not significantly influence half-cell potentials.

Acknowledgments

This study was supported by the Office of Engineering Research and the School of Civil Engineering at Oklahoma State University. Wax beads were furnished by the Oklahoma Department of Transportation.

References

[*1*] Clear, K. C. and Forster, S. W., "Internally Sealed Concrete: Material Characterization and Heat Treating Studies," FHWA Report FHWA-RD-77-16, Federal Highway Administration, Washington, D.C., 1977.

[*2*] U.S. Department of Transportation, "Internally Sealed Concrete: Guide to Construction and Heat Treatment," FHWA Report IP-77-9, Federal Highway Administration, Washington, D.C., 1977.

[*3*] Jenkins, G. H. and Butler, J. M., "Internally Sealed Concrete," FHWA Report FHWA-RD-75-20, Federal Highway Administration, Washington, D.C., 1975.

[*4*] Kansas Department of Transportation, "Rapid *In Situ* Determination of Chloride Ion in Portland Cement Concrete Bridge Decks," FHWA Report FHWA-KS-RD-75-2, Kansas Department of Transportation, Kansas City, Kans., 1976.

[*5*] Clear, K. C., "Time-to-Corrosion of Reinforcing Steel in Concrete Slabs," FHWA Report FHWA-RD-76-70, Federal Highway Administration, Washington, D.C., 1976.

[*6*] Stratfull, R. F., Jurkovich, W. J., and Spellman, D. L., "Corrosion Testing of Bridge Decks," Transportation Research Board, National Academy of Sciences, Transportation Research Record No. 539, 1975.

[*7*] Browne, R. D., Domone, P. L., and Geoghegan, M. P. in *Proceedings*, 9th Annual Offshore Technology Conference, Houston, Tex., 2–5 May 1977.

[*8*] Woods, H., *Durability of Concrete Construction*, Monograph No. 4, American Concrete Institute, 1968.

K. W. J. Treadaway, [1] *B. L. Brown,* [1] *and R. N. Cox* [1]

Durability of Galvanized Steel in Concrete

REFERENCE: Treadaway, K. W. J., Brown, B. L., and Cox, R. N., **"Durability of Galvanized Steel in Concrete,"** *Corrosion of Reinforcing Steel in Concrete, ASTM STP 713,* D. E. Tonini and J. M. Gaidis, Eds., American Society for Testing and Materials, 1980, pp. 102-131.

ABSTRACT: Two major exposure site programs have been undertaken to provide comparative information on the corrosion susceptibility of steel in concrete. In the first (current for 14 years), a comparative examination of the performance of zinc-coated and uncoated mild steel in a range of concretes made with both dense and lightweight pulverized fuel ash (pfa) aggregates has been made. In the second (which has been underway for five years and where corrosion has been accelerated by the addition of calcium chloride to the concrete), evaluation of the performance of galvanized steel (in this case on a substrate of high-yield bar) has been supplemented by studies of American Iron and Steel Institute (AISI) Types 405, 430, 302, 315, and 316 stainless steel bar and a comparison with the performance obtained from high-yield deformed bar. Both programs have employed exposure of small reinforced concrete prisms in which the concrete cover to the reinforcement has been carefully set by locating the test specimens on a supporting frame. In the five-year tests, the prism specimens have been augmented with trials on reinforced beams which have been stressed to give cracks in the concrete cover to the steel.

This paper discusses the results obtained from the galvanized steel specimens and compares them with those for the untreated steel examined in these studies. The results can be divided into a number of categories. Similar good performance (in terms of cracking of the cover induced by expansive corrosion of the reinforcement) has been exhibited by mild, high-yield, and galvanized steel in dense-aggregate good-quality chloride-free concrete. Where dense aggregate has been substituted by lightweight aggregate, cracking due to corrosion of the unprotected steel occurred at low cover, whereas identical prisms containing galvanized bars remained uncracked. The addition of high levels of calcium chloride (3.0 percent by weight and above with respect to the cement) to the dense-aggregate concrete caused severe corrosion of the high-yield bar and resulted in massive cracking of the cover. To date, this cracking has been less severe in similar specimens reinforced with galvanized steel, but results of weight loss measurement indicate extensive zinc loss in some specimens. However, much less loss of zinc has been measured on bars removed from concrete made without deliberate addition of chloride and from concrete to which up to 1.5 percent calcium chloride (by weight of cement) has been added. In the more permeable mixed without added chloride, where carbonation has reached the test bar, plain steel has corroded, resulting in cracking of the cover, whereas with

[1] Section head and research scientists, respectively, Building Research Establishment, Department of the Environment, Garston, Herts, U.K.

galvanized bar some zinc loss has occurred but without fracture of the cover. The results suggest that, although some delay in cracking of the cover is achieved by the use of galvanized reinforcement, the greatest benefit would occur where it has been used in low-quality relatively permeable concrete containing low or minimal quantities of chloride.

KEY WORDS: corrosion, galvanizing, zinc concrete rebar

Over recent years corrosion of steel in concrete has become an increasingly important factor when considering the maintenance of buildings and structures. The occurrence of this corrosion has been associated with poor-quality concrete (often in conjunction with low cover), with concrete which has been subjected to chloride ion penetration (as in marine environments, or where exposed to the use of chloride-bearing deicing salts) or to excessive additions of chloride-bearing set accelerators. In many cases the results of such corrosion are asthetically unacceptable in terms of rust staining and spalling; in some cases structural weakening (if appropriate early remedial action is not taken) occurs, and often the costs of repair are high. A number of remedies have been suggested to alleviate such problems both during design and construction and when they subsequently become apparent and represent a maintenance problem. Among remedies suggested for the design/construction phase, the application of hot-dipped galvanized reinforcement (for both buildings and civil engineering structures) has been popular. A number of examples of buildings in which galvanized reinforcement has been used can be cited; for example, in rural, urban, and industrial locations [*1*],[2] in marine exposure conditions [*2*], and for concrete subject to deicing salts [*3*].

Although documented evidence for the good performance of galvanized steel reinforcement has been published [*2,3*], controversy regarding its durability is evident in the literature and research on the topic is very active. Research carried out on the durability of galvanized steel reinforcement in concrete falls into two broad categories: laboratory studies which have, in the main, encompassed fundamental electrochemical studies of the corrosion of zinc in aqueous hydroxide environments, and exposure site tests where galvanized steel has been used to reinforce concrete test specimens in aggressive industrial and marine conditions.

Laboratory electrochemical studies carried out by Bird [*4*] considered the influence of chromate additions to the concrete on the behavior of galvanized steel with particular reference to bond strength. He concluded that the presence of small quantities (0.0035 percent) in the cement was sufficient to depress the zincate/hydrogen reaction and so improve the subsequent bond. Further electrochemical studies on the zinc/hydroxide/chromate system by Everett and Treadaway [*5*] confirmed this effect. Bird and Strauss [*6*] carried out laboratory and exposure site trials on a number of metallic coating systems for reinforcing steel; they indicated that while zinc coatings on steel

[2] The italic numbers in brackets refer to the list of references appended to this paper.

delayed the onset of corrosion, this delay was finite, especially in the presence of significant amounts of chloride, with protection afforded by cadmium proving the most satisfactory of the coating systems examined. Boyd and Tripler [*7*], using a number of electrochemical techniques, investigated the performance of steel in concrete under various conditions and included evaluation of the effect of inhibitors and metallic coatings. They concluded that for application in new concrete structures both zinc and nickel appeared suitable as possible coatings. Duval and Arliguie [*8*] used potentiodynamic polarization techniques to examine the corrosion of zinc in hydroxide/chloride solutions and concluded from results of their work that chloride attack on zinc was considerably reduced in the presence of hydroxide. Unz [*9*], in recent studies, examined the behavior of zinc and galvanized steel in chloride-bearing calcium hydroxide solutions using full- and partial-immersion techniques. The results of these tests and similar ones on mild steel were compared and it was concluded that under conditions of inhomogeneous embedment in concrete the performance of galvanized steel was questionable.

Results from exposure site testing have been similarly inconclusive. Griffen [*10*], in a series of tests in marine locations, reported questionable gain in performance from zinc-coated steel in comparison with that provided by mild steel. Baker et al [*11*], on the other hand, in a study of zinc and nickel metallic coating systems for reinforcing steel to be used in concrete exposed in marine locations, found that they did have beneficial effect in delaying cracking. This was particularly the case for nickel coatings, but zinc coatings also delayed the onset of cracking. Okamura and Hisamatsu [*12*] also examined the performance of zinc-coated reinforcing steel. In their experiments precracked beams were exposed to intermittent salt-spray conditions and after exposure were tested by fatigue and then crushed to reveal the reinforcement. The results of these experiments, which compared the performance of black and galvanized bar, suggested that a zinc coating was measureably beneficial in arresting corrosion in concrete containing cracks of up to 0.3-mm width in comparison with plain bar. Bird and Callaghan [*13*] in recent work which followed a series of long-term exposure programs reported that galvanized coatings do delay the onset of corrosion in marine environments, but do not prevent it completely. They attributed the delay to:

1. Higher salt toleration by galvanized coatings,
2. Lower efficiency of passivated zinc surfaces as cathodic oxygen reduction areas,
3. Sacrificial protection of exposed steel (authors' note: this applies to cracked concrete) once sufficient salt had penetrated, and
4. Possible inhibitive effect of zinc corrosion products on the corrosion of steel.

They continue, however:

> Galvanising is at present either used extensively or generally advocated as a solution to the problem. This is not in line with results obtained under practical conditions in South Africa. To avoid future disappointments it is felt that the time has now arrived to establish whether in fact galvanising is the answer.

It would seem, therefore, that as with the laboratory electrochemical studies some conflict in the results of exposure site studies is apparent. While studies of structures [*2*] and of bridges reinforced with galvanized steel [*14*] have suggested that galvanized bars perform well in practice, this is not necessarily reflected in exposure site results [*10,13*].

The two series of exposure site programs described herein have been designed to provide further information on the performance of hot-dipped galvanized bar in concrete. The main objectives of these two programs have been:

1. To provide information on galvanized bar in good-quality chloride-free concrete,
2. To determine the performance of galvanized bar in concrete of poor quality or in concrete subject to considerable carbonation,
3. To evaluate the effect of chloride on the performance of galvanized bar, and
4. To measure the influence of cracking in concrete on galvanized bar.

To achieve these objectives the first series of concrete specimens (which were made originally without deliberate addition of chloride) has been under test for 14 years. The second series (made up of concrete beams and prisms) has been under test for five years, but in these specimens a range of calcium chloride additions has been made to accelerate the corrosive effect.

Experimental

The exposure site studies have involved two types of reinforced concrete prisms (Series I and Series II, Fig. 1) and reinforced beams (Fig. 2) which have been stressed back to back to induce tensile cracking in the faces covering the reinforcement. Both Series I and II program specimens were exposed at Beckton, East London (industrial atmosphere) and at Hurst Castle, 6 m above the Solent (marine exposure). Additions of calcium chloride[3] were made to the Series II prisms and beams to stimulate accelerated corrosion, whereas no additions of calcium chloride were made to the Series I prisms or to the Series II beams exposed at the marine site. Table 1 provides basic details on the prisms and Table 2 details on the beams.

[3] Expressed in the remainder of the text as additions of chloride ion as a percentage by weight of cement.

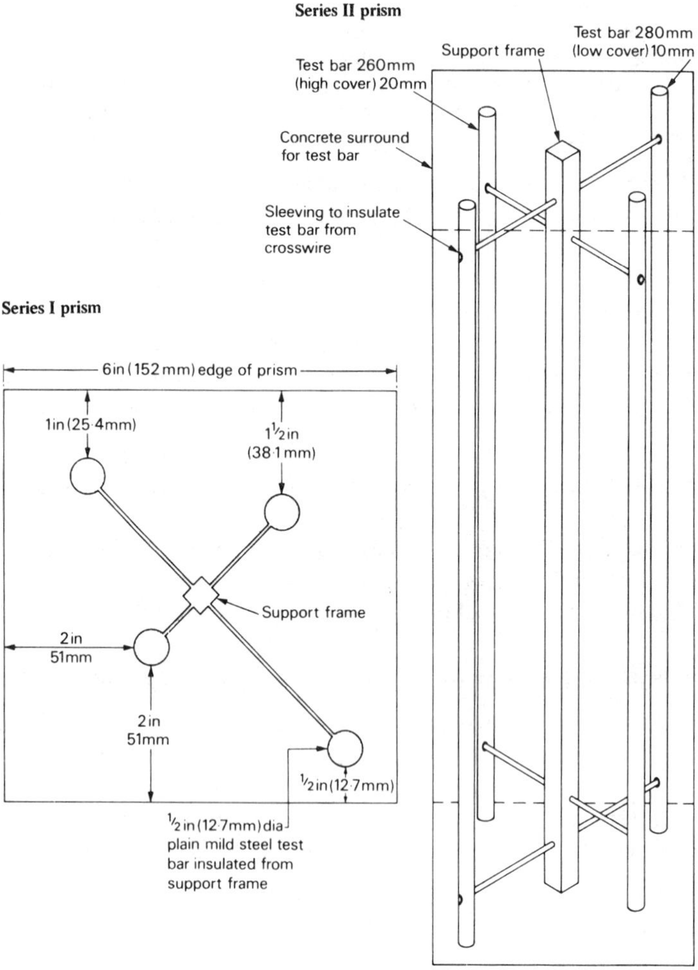

FIG. 1—*Arrangement of steel on support frame for Series I and II prisms.*

The basic design and preparation of both Series I and II prisms were similar. In both cases a central mild steel spine supported the test bars at fixed depths of cover. Series I prisms always contained bars of a single type (that is, bright mild or rusted mild steel or galvanized mild steel) with no variation in surface treatment. Series II specimens similarly contained bars of the same type, but the surface treatments varied. Thus, a prism would contain two pairs of bars with different surface treatments (shot-blasted and prerusted or plain hot-dipped galvanized and chromated hot-dipped galvanized), one being set at 10-mm cover and the other at 20-mm cover. Thus the number of variables was increased. Originally the plan for both ex-

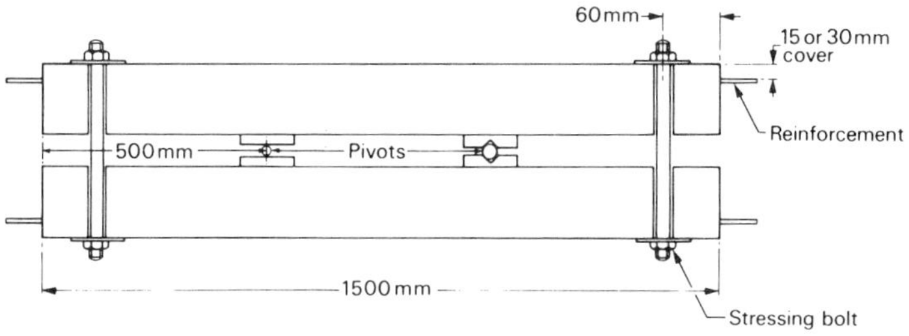

FIG. 2—*Assembly of stressed beams.*

TABLE 1—*Prism Specimens.*

	Series I	Series II
	Tests Commenced 1964	Tests Commenced 1972
Overall size	6 × 6 × 12 in. (152 × 152 × 304 mm)	100 × 100 × 300 mm
Cover to steel	½, 1, 1½, 2 in. (12.7, 25.4, 38.1 and 50.8 mm)	10 and 20 mm
Mix design/W/C[a]/28-day strength		
Dense aggregate (cement: fine aggregate: coarse aggregate) (Ham River gravel)	1:1.5:3.0/0.5, 1:2.0:4.0/0.6, 1:3.5:4.5/0.9 } /21[b]	1:2.4:3.6/0.6/39-43 1:3.2:4.8/0.75/27-31
Lightweight aggregate	1:2.02:2.02/1.0/16[b]	none
Chloride additions (% with respect to weight of cement)	none	0, 0.32, 0.96, 1.9, 3.2
Reinforcement/pretreatment	round mild steel/bright or prerusted hot-dipped galvanized (mean coating 89 μm)	deformed high-yield steel/shot-blasted prerusted hot-dipped galvanized (mean coating 88 μm) chromated hot-dipped galvanized

[a] W/C = water/cement.
[b] Design strength (units MNm^{-2}).

posure series was to remove prisms at fixed intervals of time. This was commenced for Series I prisms but it was soon discovered that as corrosion was proceeding at such a slow rate it was necessary to lengthen the original time scale to ensure that at least a few of the bars in the prisms corroded. A schedule was, however, adopted for the Series II prisms where the addition of chloride caused considerable acceleration of the corrosion process. Duplicate Series II prisms were examined at 1, 2, 3, and 5 years, and a pair of prisms

TABLE 2—*Beam specimens (Series II exposure program).*

Commencement of Tests	1971 (Pairs of Beams Being Inverted Periodically)
Low cover	1500 × 100 × 150 mm
High cover	1500 × 100 × 165 mm
Cover to steel	15 and 30 mm
Mix design/W/C[a]/28 day strength	
Dense aggregate	1:2.0:3.0/0.55/50-51 MN/m^2
Chloride additions (% with respect to cement)	0, 0.32, 0.96, 3.2
Reinforcement/pretreatment	deformed high-yield steel/shotblasted or prerusted hot-dipped galvanized (mean coating thickness 88 μm) chromated hot-dipped galvanized

[a] W/C = water/cement.

from each group will remain on exposure to be examined 10 years after casting. For both series, prior to embedment, the preweighed bars were assembled onto the metal support frame (Fig. 1), electronic contact between frame and bar being avoided by using an insulating sleeve. During casting this assembly was held in position with a socket fixed in the base of the mold to ensure its correct location.

The Series II beam specimens were designed to provide 15 or 30 mm cover. After casting and curing, the beams were placed on exposure and cracked by fastening pairs of beams of similar cover back to back across roller pivots (Fig. 2). Stressing bolts, passing through holes at each end of the beams, then were tightened slowly to provide stresses in the face of the concrete necessary to produce cracks normal to the reinforcing bar with widths of up to 300 μm at the beam surface. Table 2 indicates that amounts of chloride, similar to those added to the Series II prisms, were added to the beams, and exposure durations also were similar to those for the Series II prisms.

Both series of specimens have been regularly inspected at the exposure sites. Cracks associated with the corrosion of the reinforcement bars have been noted as they appeared, and in more extreme cases photographic records of such features have been made.

On recovery, the specimens were broken open in the laboratory. The extent of carbonation of the prisms was assessed by spraying the fresh fractures with phenolphthalein solution and, where appropriate, the strength of the concrete was checked by crushing test cubes. The reinforcement was carefully removed from the specimens and then examined for rusting and other corrosion effects. Appropriate cleaning solutions were used to remove corrosion products, and the cleaned test bars from the prisms were weighed to provide a measure of the corrosion. After cleaning, a visual examination was made of the reinforcement from both the beams and the prisms for pitting and other corrosion effects.

Results

Three main measures of corrosion performance of the test steels have been used in this study: (1) cracking of the cover associated with formation of expansive corrosion products on the test bars, (2) weight loss on the steel, and (3) condition of the steel. The first of these measures was nondestructive and relied on periodic examination of the specimens, whereas the latter two both required destruction of the specimen prior to measurement or examination of the steel.

Cracking

The information obtained from periodic examination of both Series I and II prisms is given in Tables 3 and 4. It will be seen from Table 3 that in dense-aggregate concrete cracking has occurred only where the leanest mix has been used, and only with prerusted steel set at ½-in. (12.7 mm) cover. The incidence of cracking is much higher in the case of lightweight aggregate concrete. Cracking at ½-in. cover was observed in all the lightweight-aggregate prisms reinforced with bright or prerusted mild steel when examined in July 1970 (although in the case of four prisms reinforced with bright bar, cracking was not apparent during the June 1970 inspection). It is important to note that at the most recent examination (May 1978) none of the prisms reinforced with galvanized bar (including the lightweight-aggregate concrete) showed cracking.

The Series II prisms also have been examined periodically and Table 4 provides information on the time of first observation of cracking and, where applicable, the time taken for half and all of specific test bars to exhibit cracking of the cover. Some obvious features appear from inspection of this information. It seems that for high-yield bar the incidence of cracking increases as the percentage of chloride added to the concrete increases, and, as the chloride in the concrete increases, cracking is observed at earlier exposure times. Indeed, there is a readily discernable increase in cracking as the chloride ion content is raised from 0.96 to 1.9 percent by weight of cement. It seems also that corrosion on prerusted bars occurring after embedment causes cracking at slightly earlier ages than similar corrosion on shotblasted bars. Furthermore, increasing the cement content and depth of cover from 10 to 20 mm delays the onset of corrosion. While a similar trend in performance can be noted when high-yield bar is replaced by galvanized bar under equivalent conditions, a delay in the onset of cracking and some amelioration of cracking incidence are apparent. This is especially the case where up to 0.96 percent chloride ion has been added to the cement, although even at an addition of 1.9 percent some reduction in cracking is apparent. However, cracking incidence is still high where 1.9 or 3.2 percent chloride additions have been made.

TABLE 3—*Series I prisms: inspections for cracks.*

		Bright Mild Steel		Prerusted Mild Steel		Galvanized Steel	
Concrete Type	Inspection Date	Uncracked	Cracked	Uncracked	Cracked	Uncracked	Cracked
Dense 1:1.5:3	July 1970	7	0	4	0	10	0
	May 1978	7	0	4	0	10	0
Dense 1:2:4	July 1970	12	0	9	0	17	0
	May 1978	12	0	9	0	17	0
Dense 1:3.5:4.5	July 1970	14	0	1	4	11	0
	May 1978	14	0	0	5	11	0
Lightweight 1:2.02:2.02	July 1970	0*	12	0	6	16	0
	May 1978	0	12	0	6	16	0

NOTES:

Data refer to numbers of prisms either cracked or uncracked.

Cracking has been observed only on bars embedded at ½-in. cover.

*Indicates that four of the prisms which were cracked in July 1970 were not cracked in June 1970.

TABLE 4—*Series II prisms: inspections for cracks.*

Mix	Chloride, %	Pretreatment[a]	Cover, mm	Exposure Time, days		
				First Crack	50% Cracked	100% Cracked
			High-Yield Steel			
1:6	0.0	R	10	256	1546	...
	0.0	SB	10	597	...	...
	0.32	R	10	256	1173	...
	0.32	R	10	486	...	...
	0.96	R	10	154	154	...
	0.96	SB	20	486	...	...
	1.9	R	20	23	109	730
	1.9	R	10	23	154	710
	1.9	SB	10	109	175	...
	1.9	SB	20	109	456	...
	3.2	R	10	23	23	175
	3.2	R	20	23	109	1546
	3.2	SB	10	23	109	1546
	3.2	SB	20	109	197	1546
1:8	0.0	SB	20	338	730	...
	0.0	SB	10	486	...	...
	0.0	R	10	1220	...	...
	0.32	R	10	256	456	...
	0.32	SB	10	1220	...	...
	0.96	R	10	154	175	...
	0.96	R	20	154	197	1220
	0.96	SB	10	154	1220	...
	0.96	SB	20	197	1172	...
	1.9	R	10	23	154	197
	1.9	SB	10	23	197	535
	1.9	R	20	154	154	535
	1.9	SB	20	197	1220	...
	3.2	R	10	109	109	175
	3.2	R	20	109	109	197
	3.2	SB	10	154	175	197
	3.2	SB	20	154	197	1173
			Galvanized Steel			
1:6	0.0	AR	20	23[b]	...	...
	0.32	CHR	10	332	...	...
	0.32	AR	10	1220	...	...
	0.96	CHR	20	294	...	...
	0.96	AR	10	338	1173	...
	0.96	CHR	10	338	1173	...
	1.9	AR	10	338	1173	...
	1.9	CHR	10	338	1173	...
	3.2	AR	10	197	256	256
	3.2	CHR	10	256	256	456
	3.2	CHR	20	294	...	...
1:8	0.0	CHR	10	1546	...	...
	0.96	CHR	20	456	...	...
	1.9	CHR	10	294	1937	...
	1.9	AR	10	294	...	...
	3.2	CHR	10	233	256	294
	3.2	AR	10	233	294	456
	3.2	AR	20	233	535	1546
	3.2	CHR	20	256	535	...

[a] R = prerusted; SB = shotblasted; AR = as received; CHR = chromated.
[b] Insignificant weight loss recorded on bar associated with crack on recovery.

The "in-service" performance of the Series II beams has been followed in a similar manner to that of the Series II prisms, regular inspections having been made. Initially the positions of tensile cracks on the stressed beams surfaces were mapped and their widths measured, the results falling in the range 25 to 250 μm, with wider cracks formed midway along the beams. As exposure progressed cracks became blocked (healed), with products of carbonation and debris washed from the concrete surface. This coupled with degradation of the crack edges made crack width measurement increasingly difficult and as a result this aspect of the study was discontinued.

After 3½ years' exposure at the industrial site, longitudinal cracking associated with reinforcement corrosion was observed in beams reinforced with high-yield steel and to which 3.2 percent chloride had been added. Similar cracking was observed on beams reinforced with galvanized bar after four years and on beams reinforced with high-yield steel exposed for five years at the marine site (no chloride had been added to this concrete), identical beams reinforced with galvanized steel have not, as yet, shown evidence of similar cracking. As no chloride additions were made to concrete tested at the marine site, the occurrence of cracking associated with reinforcement corrosion gives an indication of the relative severity of the exposure conditions in comparison with those at the industrial site. It would seem that, for cracked concrete, initially chloride-free marine conditions are more severe than where 0.96 percent chloride ion is added to cracked concrete which is subsequently exposed at the industrial site.

Weight Loss

The weight loss data obtained from Series I and II prisms have been used to assess (see Appendix I) the corrosion rates of the steel and galvanized steel test bars in the range of concretes examined.

Series I—Data obtained on the corrosion rates of bars in the Series I prisms are presented in Table 5.

Bright mild steel bars embedded in dense-aggregate concrete have performed well over the test period. The depth of penetration of the carbonation front ranged from 1 to 10 mm, the steel loss not exceeding 1 μm, which was insufficient to cause cracking of the cover. The corrosion loss on the prerusted steels in dense-aggregate concrete, however, was much higher, the range, in uncracked concrete where carbonation had not reached the minimum depth of cover, falling between 29 and 45 μm (after adjustment for the original prerust), and reaching 90 μm where the ½-in. cover had fully carbonated and subsequently cracked. Corrosion of the prerusted steel was of an overall nature rather than exhibiting specific sites of attack. In the lightweight concrete cracking had occurred over all the ½-in. cover mild steel bars after approximately six years' exposure, and after 14 years this had been translated to spalling in several cases. Carbonation of the cover after 14 years

TABLE 5—*Thickness loss on bars recovered from Series I prisms.*

Steel/Concrete Type	Carbonation, mm	Metal Lost (μm) in Relation to Cover			
		½-in.	1-in.	1½-in.	2-in.
Prerusted/					
Dense 1:1.5:3	2-8	31	32	45	31
Dense 1:3.5:4.5	8-12	90	29	32	30
Lightweight	12	155[a]	28	30	31
	16	195[a]	27	30	29
	19	167[a]	29	31	30
Bright/Dense 1:2:4	2-10	1	1	1	1
	0-3	1	1	1	1
	1-5	1	1	1	1
Dense 1:3.5:4.5	1-8	1	1	1	1
Lightweight	12 at low cover 24 elsewhere	12[b]	4	1	1
	10	27[a]	3	1	1
	13	23[a]	2	1	1
Galvanized/					
Dense 1:1.5:3.0	1-2	1	1	1	1
	2-3	1	1	1	1
	2-6	1	1	1	1
Dense 1:2:4	2-4	1	1	1	1
	2-8	2	1	1	1
	1-8	1	1	1	1
	3-8	1	1	1	3
	2-6	1	1	2	1
Dense 1:3.5:4.5	10-12	6	3	2	3
	13	4	1	2	2
	13	4	2	2	2
Lightweight	12	7	3	4	2
	16	9	2	2	2
	25	10	2	3	2
	22	8	3	3	3
	22	11	10	3	3
	24	10	3	4	4

[a] Cracked and spalled at low cover.
[b] Cracked at low cover.

All data are given to the nearest whole number. Dense concrete prisms containing bright and prerusted steel were removed from exposure site after 12 years' exposure. Dense and lightweight aggregate prisms containing galvanized steel were removed from exposure after 14 years. Lightweight aggregate prisms containing bright and prerusted steel were removed from exposure after 14 years. Above data collected from prisms exposed at the industrial exposure site.

1 in. = 25.4 mm.

ranged from 10 to 24 mm and on removal from the concrete all ½-in. cover bars were rusted and pitted to some extent. Corrosion of the prerusted bars was severe, with loss between 155 and 195 μm at the lowest cover. In comparison, bright bars at the same cover lost between 12 and 27 μm. For both bright and prerusted bars losses at deeper cover decreased although there was still a substantial difference between the two initial surface conditions.

However, none of the corrosion was sufficient to create cracking at the higher depths of cover.

Galvanized bars have also performed very well in the dense-aggregate concretes. The range of carbonation depths has fallen between 1 and 13 mm, the greatest penetration being observed in the leanest mix; the zinc loss appears to increase with increasing carbonation and reduced cement content. It is important to note that to date none of these prisms has exhibited cracking.

Carbonation has been high in lightweight-aggregate concrete, ranging from 16 to 25 mm. In all cases the zinc loss has been much higher at ½-in. cover in lightweight concrete than has been the case in dense-aggregate concrete at similar cover, and there appears to be a similar but far less marked trend to increased zinc losses at the other depths of cover. It is important to note, however, that none of the prisms have cracked parallel to the steel, and in terms of total zinc thickness only about 10 percent has been lost during the 14 years' exposure. When removed from uncarbonated dense-aggregate concrete, the galvanized bars showed only slight attack. In carbonated concrete at low cover, some discontinuities in the zinc coatings were apparent and these were accompanied by a few rust spots. In carbonated lightweight-aggregate concrete some slight red-rusting was apparent but in uncarbonated material the zinc was still in good condition.

Series II—The weight loss data converted to thickness loss for bars removed from Series II prisms are given in Table 6.

For shotblasted high-yield steel, thickness loss on exposure showed, in both 1:6 and 1:8 cement:aggregate concretes, a sharp increase when 1.9 percent or more chloride had been added as a percentage of the cement to the concrete. Considerable loss (20, 24 μm) was observed also after five years' exposure in 1:8 concrete at low cover with 0.96 percent added chloride. Substitution by prerusted steel increased thickness loss (after compensation for initial prerust) in comparison with shotblasted steel exposed in similar circumstances. In concrete made without added chloride, losses of up to 14 μm occurred on shotblasted bars, whereas losses were slightly higher on prerusted steel.

Thickness losses on galvanized reinforcement were small at low chloride additions, but loss of zinc was extensive where 1.9 percent or more chloride had been added. In these cases breaks in the zinc coating had led to severe rusting of the underlying steel, and in many cases the greater part of the zinc had been lost.

Visual examination of steel removed from concrete containing high chloride levels revealed severe corrosion and pitting on prerusted bars, and in some of these very significant cross-sectional loss. Similar but less-extensive attack was observed on shotblasted bars in identical concretes. At high chloride levels galvanized bars had also suffered severe corrosion, but it was less pronounced than on either the prerusted or shotblasted bars, Fig. 3 illustrating these effects. At lower chloride levels (0.96 percent and below) corrosion was less severe although pitting was occasionally apparent on the prerusted and shotblasted bars. Galvanized steel showed an increase in protection as illustrated in Fig. 4.

Examination of Steel from Beams

A few general remarks can be made that are applicable to both galvanized and high-yield bars removed from beams after exposure. After breaking the bars from the concrete, it was observed that an area of bond breakdown (accompanied by loss of alkalinity at the steel/concrete interface) occurred at the intersection of the tensile crack and the reinforcement. The greatest area of bond breakdown was on the surface of the bar closest to the tensile face of the concrete, but in many cases a lesser amount of breakdown had occurred on the inner face. These areas of bond disruption varied and did not appear to have any relationship to crack width or concrete cover. It was in these areas that the majority of corrosive attack took place (see Table 7).

The degree of loss of alkalinity measured in the concrete away from the influence of cracking did not exceed 3 mm.

Galvanized Steel Bars

Only superficial corrosion was noted on galvanized bars which had been embedded in concrete containing up to (and including) 0.96 percent chloride exposed at the industrial site for two years. Similar observations were made on bars removed from the beams that had been exposed for the same period at the marine site. In all these cases corrosion was noted only at the area of intersection of the tensile crack and the bar and in the area of bond breakdown. After three years' exposure, however, bars embedded at high cover in concrete containing 0.96 percent chloride also showed corrosive attack in areas divorced from the tensile cracks.

After five years' exposure at the industrial site, where no added chloride was present in the concrete, the corrosion was still restricted to the areas in the vicinity of the tensile cracks. The degree of corrosion was greater than after two years but the zinc had not been breached. There was some white rust and blackening of the zinc, however. Where chloride had been added, corrosion was also occurring at areas divorced from the cracks; the extent of corrosion increasing with increasing chloride content. Where 3.2 percent chloride was added, the zinc had been lost in some areas, accompanied by corrosion of the underlying steel. This has resulted in longitudinal cracking of the concrete cover which originated from the buildup of corrosion product initiated at a tensile crack.

The beams exposed for five years at the marine site have suffered corrosion of the zinc with the formation of white rust and some areas of blackening, but, again, this has been associated with original tensile cracking.

TABLE 6—*Mean thickness loss (μm) on bars removed from Series II prisms.*

Years on Exposure	1	2	3	5	1	2	3	5	1	2	3	5	1	2	3	5
% Added Cl^- (by weight cement)	High-Yield Steel				Shortblasted				High-Yield Steel				Prerusted			
	1:6 Concrete Mix		High Cover		1:8 Concrete Mix		High Cover		1:6 Concrete Mix		High Cover		1:8 Concrete Mix		High Cover	
0	2	10	4	9	3	4	3	7	6	12	20	16	3	17	2	12
	3	11	4	11	2	5	4	7	7	2	7	9	6	10	16	13
0.32	1	10	3	9	1	6	3	8	10	30	19	22	9	2	17	21
	5	4	9	8	1	9	3	10	6	17	1	19	*a*	16	*a*	1
0.96	1	5	4	8	4	10	10	10	13	15	11	36	33	23	11	50
	1	6	4	7	2	10	9	12	5	2	5	4	17	53	38	87
1.9	5	22	13	15	15	23	28	51	31	40	21	174	42	54	36	223
	5	20	10	17	15	29	30	58	10	28	145	151	81	85	66	82
3.2	22	29	23	50	26	35	46	75	108	90	234	377	17	121	245	462
	17	26	31	55	28	39	46	76	19	109	80	508	57	69	218	145
	1:6 Concrete Mix		Low Cover		1:8 Concrete Mix		Low Cover		1:6 Concrete Mix		Low Cover		1:8 Concrete Mix		Low Cover	
0	29	11	5	8	3	7	4	14	*a*	*a*	*a*	8	8	16	8	11
	3	11	4	9	3	4	7	11	10	20	3	19	16	20	9	23
0.32	3	11	4	2	2	9	4	7	*a*	13	*a*	20	1	14	22	51
	3	4	3	7	1	11	4	6	*a*	6	9	13	3	9	18	25
0.96	1	6	3	14	10	12	13	20	4	34	13	15	12	51	33	75
	2	5	6	11	6	18	15	24	16	9	14	17	11	36	104	15
1.9	6	25	12	19	15	24	31	40	14	50	14	133	90	109	102	49
	7	23	22	20	18	26	34	34	68	106	50	58	66	36	201	273
3.2	35	42	29	71	30	41	43	57	197	55	32	141	58	71	84	119
	29	30	52	49	32	42	47	74	121	84	229	267	18	93	170	222

	Galvanized Steel				As Received				Galvanized Steel				Chromated			
	1:6 Concrete Mix		High Cover		1:8 Concrete Mix		High Cover		1:6 Concrete Mix		High Cover		1:8 Concrete Mix		High Cover	
0	7	2	2	12	2	1	2	14	8	4	3	12	3	1	2	13
	2	2	3	9	2	2	1	15	2	2	4	12	2	3	2	14
0.32	2	1	1	10	1	1	2	11	2	1	3	10	2	2	2	9
	1	5	2	10	1	1	1	9	1	3	3	9	1	1	1	11
0.96	1	2	2	2	1	1	2	5	2	2	2	2	1	1	1	2
	3	2	12	2	1	1	2	2	5	1	2	2	2	1	2	2
1.9	29	22	24	*b*	20	26	44	43	17	6	8	58	6	6	29	41
	36	27	27	30	27	28	34	*b*	13	1	5	12	6	17	24	53
3.2	*b*	72	6	*b*	*b*	*b*	77	*b*	*b*	48	53	*b*	*b*	*b*	55	*b*
	b	73	69	*b*	63	*b*	68	*b*	54	47	51	60	38	*b*	54	*b*
	1:6 Concrete Mix		Low Cover		1:8 Concrete Mix		Low Cover		1:6 Concrete Mix		Low Cover		1:8 Concrete Mix		Low Cover	
0	12	3	3	10	1	1	2	6	14	3	4	17	1	1	2	5
	1	1	4	13	1	1	1	6	2	2	5	13	1	1	1	7
0.32	2	1	2	11	1	1	1	5	3	9	2	12	1	2	1	4
	1	3	3	10	1	1	1	4	1	1	4	13	1	1	1	3
0.96	3	2	2	1	1	1	4	9	1	3	2	5	1	1	1	5
	6	4	5	2	1	11	4	3	5	1	7	2	1	1	1	3
1.9	30	31	32	*b*	27	34	36	44	28	18	17	55	12	21	24	27
	30	33	30	37	26	25	36	*b*	30	15	17	20	9	22	28	33
3.2	84	55	62	*b*	*b*	*b*	*b*	*b*	46	44	38	*b*	*b*	*b*	44	*b*
	62	63	62	*b*	48	*b*	60	*b*	37	*b*	52	*b*	35	64	57	*b*

[a] Below average prerust.
[b] Corrosion penetrated to steel substrate.

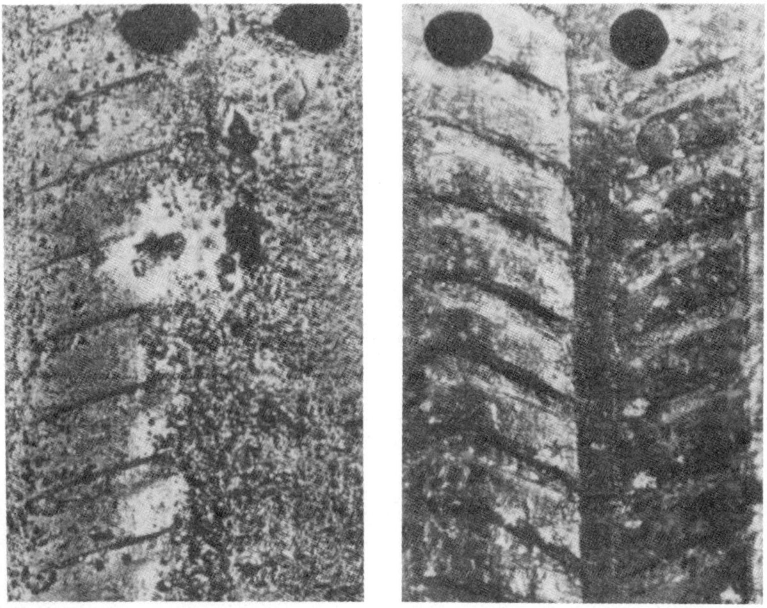

FIG. 3—*Reinforcement from Series II prisms after five years' exposure. Magnification ×1.5; pictures taken with a periphery scanning camera.*

Visual examination of galvanized bars after removal from the beams did not reveal differences in performance between bars in preembedment chromated and as-received surface conditions.

Unprotected Steel Bars

A pattern of results similar to those for the galvanized bars has been observed with the high-yield steel. In all cases, however, the degree of corrosion has been greater and longitudinal cracking has been observed at earlier times. Furthermore, while longitudinal cracking has been observed on beams reinforced with high-yield steel at the marine site, none has yet been observed where galvanized bars have reinforced beams at the same site.

Observation of the condition of the high-yield bars could not detect any difference in performance between prerusted and shotblasted bars.

Discussion

A number of objectives (assessed in relation to corrosion-associated cracking of the cover and metal loss) of the work described herein were identified in the beginning of this paper. They were

1. To provide information on the performance of galvanized bar in good-quality chloride-free concrete,

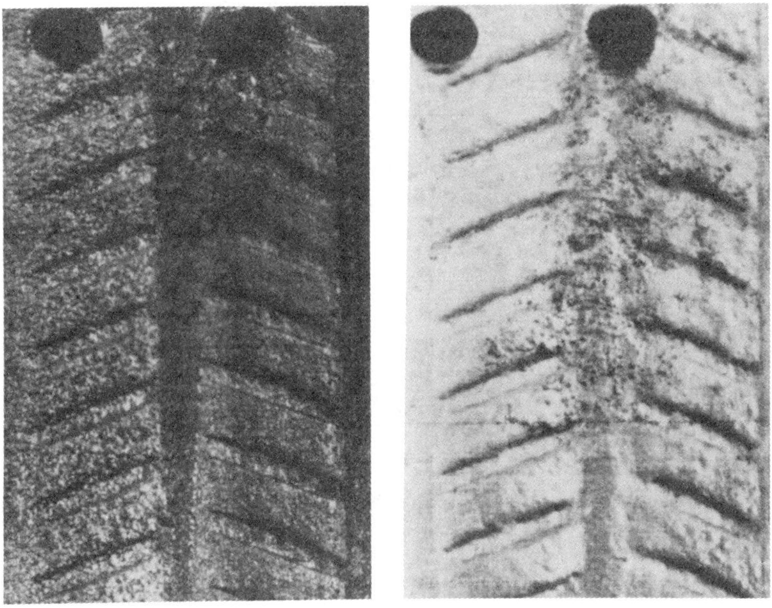

FIG. 4—*Reinforcement from Series II prisms after five years' exposure. Magnification ×1.5; picture taken with a periphery scanning camera.*

2. To determine the performance of galvanized bar in concrete of poor quality or concrete subject to considerable carbonation,
3. To evaluate the effect of chloride on the performance of galvanized bar, and
4. To assess the influence of precracking of concrete on the performance of galvanized bar.

Furthermore, the experiments were designed to compare the performance of galvanized bars and high-yield and mild steel, and to compare in each bar type the influence of surface condition, for galvanized bar as received and chromated, and for the steel rust-free and prerusted. Accordingly, these factors will be considered when discussing the results of the study.

Good-Quality Chloride-Free Concrete

In both Series I and II exposure programs bright bar and galvanized bar have performed well, this being especially the case in the Series I exposures where both bright and galvanized bar have been removed from the prisms after exposure in the almost "as embedded" condition. In these, no serious corrosion has occurred in good-quality concrete where carbonation depths have not reached the depth of cover. Series II results have been slightly more variable as a consequence of misplacement of some of the prism supports during casting, leading to less than nominal depths of cover, and in these

TABLE 7—*Results of examination of steel from Series II beams.*

Industrial Exposure Site.

Beam No.	Chloride Content	Exposure time, years	Corrosion in Crack Zone		Corrosion Divorced from Cracks		Cover	Remarks
Galvanized Steel								
			L	O	L	O		
1	0.0	2	✓	✓			H	
4	0.0	2	✓	✓			H	
5	0.0	2	✓	✓			L	
6	0.0	2	✓	✓			L	
741	0.0	5	✓	✓			H	
743	0.0	5	✓	✓			H	
742	0.0	5	✓	✓			L	
744	0.0	5	✓	✓			L	
36	0.32	2	✓	✓		✓	H	
35	0.32	2	✓	✓			H	
33	0.32	2	✓	✓			L	
35	0.32	2	✓	✓		•	L	
37	0.32	5	✓	✓	✓	✓	H	
38	0.32	5	✓	✓	✓	✓	H	
39	0.32	5	✓	✓	✓	✓	L	
40	0.32	5	✓	✓	✓	✓	L	
41	0.92	3	✓	✓	✓	✓	H	
43	0.92	3	✓	✓	✓	✓	H	
42	0.92	2	✓	✓		✓	L	
44	0.92	2	✓	✓			L	
45	0.92	5	✓	✓	✓	✓	H	
46	0.92	5	✓	✓	✓	✓	H	
47	0.92	5	✓	✓	✓	✓	L	
48	0.92	5	✓	✓		✓	L	
147	3.2	5	✓	✓	✓	✓	H	longitudinal cracking of concrete beam
148	3.2	5	✓	✓	✓	✓	H	
145	3.2	5	✓	✓	✓	✓	L	
146	3.2	5	✓	✓	✓	✓	L	
Unprotected Steel								
			L′	O′	L′	O′		
737	0.0	2	✓	✓			H	
739	0.0	2	✓	✓			H	
738	0.0	2	✓	✓			L	
740	0.0	2	✓	✓			L	
746	0.0	5	✓	✓			H	
748	0.0	5	✓	✓			H	
745	0.0	5	✓	✓			L	
747	0.0	5	✓	✓			L	
19	0.32	2	✓	✓			H	
20	0.32	2	✓	✓			H	
17	0.32	2	✓	✓			L	
18	0.32	2	✓	✓			L	
21	0.32	5	✓	✓	✓	✓	H	
22	0.32	5	✓	✓	✓	✓	H	
23	0.32	5	✓	✓	✓	✓	L	

TABLE 7—*Continued.*

Industrial Exposure Site.

Beam No.	Chloride Content	Exposure time, years	Corrosion in Crack Zone		Corrosion Divorced from Cracks		Cover	Remarks
			Unprotected Steel—Continued					
24	0.32	5	✓	✓	✓	✓	L	
25	0.96	5	no data available				not removed from site	
27	0.96	5	no data available				not removed from site	
26	0.96	2	✓	✓	✓	✓	L	
28	0.96	2	✓	✓	✓	✓	L	
29	0.96	5	no data available				H	
30	0.96	5					H	
31	0.96	5					L	
32	0.96	5					L	
151	...	5	not removed from site				H	longitudinal cracking of concrete
152	...	5					H	
149	...	5	not removed from site				L	
150	...	5					L	
			Galvanized Steel					
			L″	O″				
75	0	2	✓	✓			H	
76	0	2	✓	✓			H	
73	0	2	✓	✓			I	
74	0	2	✓	✓			L	
79	0	5	✓	✓			H	
80	0	5	✓	✓			H	
77	0	5	✓	✓			L	
78	0	5	✓	✓			L	
			Unprotected Steel					
			L′	O′	L′	O′		
81	0	2	✓	✓			H	
82	0	2		✓			H	
83	0	2					L	
84	0	2	✓				L	
85	0	5	✓	✓			H	
87	0	5	✓	✓		✓	H	
86	0	5	✓	✓		✓	L	
88	0	5	✓	✓	✓		L	longitudinal cracking of concrete

NOTES:

L = chromated galvanized bar.
O = galvanized bar.
L′ = shotblasted steel.
O′ = rusty steel.
L″ = chromated galvanized steel (marine exposure).
O″ = galvanized steel (marine exposure).
L‴ = shotblasted steel (marine exposure).
O‴ = rusty steel (marine exposure).

cases (where carbonation has reached the depth of embedment) some corrosion has occurred which has led to cracking of the cover. However, where shotblasted or galvanized steel has been in uncarbonated concrete, little measureable corrosion has been observed. Prerusting has had some influence, however, in that a much higher metal loss has been measured than on the bright or shotblasted bars. It seems that some of the inhibitive properties of the cement matrix could be lost by restriction of passivating ions to the steel surface by the physical influence of the rust, and by possible incorporation within the rust scale of aggressive ions which could exert an accelerating influence on corrosion rate.

Poor-Quality Chloride-Free Concrete

The main emphasis of tests within poor-quality concrete has been in the Series I program. Both lean dense-aggregate (1 : 3.5 : 4.5 cement : sand : coarse aggregate) and lightweight-aggregate mixes have suffered cracking as a consequence of steel reinforcement corrosion, and in these examples carbonation has been a significant factor. Carbonation has reached the minimum depth of cover in all the lightweight-aggregate prisms so far examined, and where these have been reinforced with mild steel cracking of the cover following steel corrosion has occurred. It seems that cracking has appeared at an earlier stage in the exposure where the bar was prerusted, but the delay resulting from the use of bright bar has been small. Carbonation has been similar in those lightweight-aggregate prisms reinforced with galvanized bar, but in these prisms no cracking has yet occurred as a result of expansive corrosion of the reinforcement. Indeed, at the current stage in the trials (after 14 years' exposure) there has been a delay of approximately seven years between the observation of 100 percent cracking at low cover in the lightweight-aggregate prisms reinforced with mild steel bar and the appearance of any cracking at similar cover where galvanized bar has been used. Furthermore, examination of the galvanized bar has indicated it to be in good condition with approximately 10 percent loss in zinc thickness over the exposure period. This would indicate, assuming a relatively linear corrosion rate and uniform corrosion, that the galvanized steel has considerable residual protection before the steel substrate becomes corroded.

Chloride in Concrete

Deliberate additions of chloride to the concrete were made only in the Series II program, and remarks on the influence of chloride will be confined to specimens within this program.

It appears that the addition of high levels of chloride to reinforced concrete can have an overriding effect on the corrosion performance of the reinforcing steel. Present results from the Series II prism program indicate that when 1.9

percent or more chloride is added to the concrete (equivalent to 5.4 lb chloride ion per cubic yard (3.2 kg/m^3) of 1:8 cement: aggregate concrete and 7.3 lb chloride ion per cubic yard (4.3 kg/m^3) of 1:6 concrete) severe corrosion of both shotblasted and prerusted steel occurs, leading to considerable reduction in cross section of the steel and massive cracking of the concrete both at low (10 mm) and high (20 mm) cover. When galvanized bar is compared with high-yield bar exposed under equivalent conditions, some amelioration of cracking was apparent, and delays approaching a year have been found, but zinc losses over five years' exposure at these high chloride levels have been severe (in some cases approaching the total coating weight of zinc). It is likely that under such severely aggressive conditions these coatings only delay corrosion of the substrate steel.

At lower chloride levels (up to 0.96 percent) corrosion was much less severe, although some attack has become apparent on prerusted bars at low cover in concrete containing additions of 0.96 percent chloride, and this has been sufficient to cause cracking of the cover. Galvanized steel under equivalent conditions has shown little loss in zinc and its use has undoubtedly delayed the corrosion-induced cracking, in some cases up to three years.

Tensile Cracking in Concrete

The results of the Series II beam program complement observations made on the prisms regarding the influence of chloride in promoting corrosion. Increasing the chloride content of the concrete has increased the incidence and severity of corrosion of high-yield bar at its intersection with tensile cracks in the concrete matrix, and has at the highest level (3.2 percent chloride) led to corrosion elsewhere on the test bars. Simultaneously at this chloride level, sufficient corrosion has occurred on the high-yield bar to cause cracking of the cover. However, the use of a higher cement content in the beam specimens has had a marked overall influence compared with the prism specimens in apparently reducing the general severity of the corrosion, although severe attack is still apparent at the intersections of some tensile cracks with bars. Considering galvanized bar exposed in equivalent conditions to high-yield steel, as with the prisms, some amelorative effect was found. In this context it is interesting to note a delay in the occurrence of corrosion-induced cracking from 3½ to 4 years observed between high-yield and galvanized bar in concrete containing the highest level of chloride.

The influence of crack width has been difficult to quantify in the present series of tests. Although crack width measurement was abandoned at an early stage in the study, it was apparent that ranges of crack widths had developed at the beam surfaces. When the steel was examined after exposure, however, no relationship between area of steel corroded and initial crack width could be deduced. Although this appears surprising, the problem of influence of crack width on corrosion has recently been considered by

Beeby [*15*], who suggested that, while important, crack width was not an overriding factor influencing corrosion of the underlying steel.

Surface Treatment

It will be observed from the preceeding comments that, in the prism specimens, prerusting the steel appeared to promote earlier cracking of the cover, and thickness loss measurements indicated a higher rate of corrosion than on bars with a bright or shotblasted surface. The chromating applied to galvanized bars has also been a subject of this study and, while qualitative assessment of the performance of prerusted and bright steel has been possible, this has not been the case with galvanized bar. Examination of thickness loss measurements has shown little apparent difference in performance between the as-received and chromated bars in relation to cracking of the cover (although the interface between chromated bar and concrete appeared less spongy than that between as-received bar and the concrete). However, some statistical analysis of the comparative thickness loss data has been attempted (Appendix II). This has shown that a difference in performance is discernable, and that chromating provides marginally better performance in relation to thickness loss than the as-received bar. In the present series of tests this difference is small, but earlier work [*4,5*] has indicated that the degree of benefit accruing from chromating the surface of galvanized bar is dependent, to an extent, on the availability of soluble chromate in the cement. There is little doubt, however, that even in the conditions of the Series II tests, benefit has accrued from chromating, both in terms of reduction of hydrogen formation at the metal/concrete matrix interface and in reducing zinc loss, thus supporting the practice of chromating the galvanized steel prior to embedment in concrete.

General Considerations

Comparison of the data obtained from the work described in this paper can be made with results from other studies. Bird and Strauss [*6*], Bird and Callaghan [*13*], and Baker et al [*11*] all reported delay in the onset of cracking when galvanized bar was substituted for high-yield bar in concrete exposed in marine conditions. However, Bird and Strauss [*6*] and Unz [*9*] expressed reservations about the performance of galvanized bar in high-chloride conditions, reservations which are supported in the present work. Furthermore, Bird and Callaghan, in discussing a delay to onset of corrosion imparted by galvanizing the reinforcement, implied that such a coating system would have a finite life (dependent upon exposure conditions and other factors which normally control the time to corrosion of steel in concrete). This would also be a conclusion from the present study, but it would seem from this work that measurable benefit is gained in relatively chloride-free poor-quality concrete. Finally, Okamura and Hisamatsu [*12*] have observed similar beneficial effects of substitution of galvanized steel for high-yield bar in cracked concrete exposed in marine conditions similar to those observed in the present work.

Data from the present work support the thesis that the major factor in protecting steel embedded in concrete is the alkalinity which is produced during hydration of the cement compounds. Loss of this alkalinity by carbonation, or disruption of its protective ability in the presence of sufficient concentration of aggressive ions (particularly chloride ions), will lead to corrosion. When the reinforcement has suffered corrosion sufficient to produce forces (resulting from the formation of expansive corrosion products) which are greater than the tensile strength of the cover, cracking longitudinal to the corroding bar will result, and this physical manifestation has been taken as the criterion of corrosion in this paper, and will also be used in the following part of the discussion as the life required to first maintenance.

The processes which lead to corrosion cracking of reinforced concrete can be summarized as follows:

1. An initiation period in which the concrete loses its protective capabilities by carbonation, by chloride attack, or by a combination of these two factors.
2. Corrosion of the reinforcement.
3. Cracking of the cover.

In well-designed concrete construction using good-quality concrete, Stage 1 in the corrosion process should exceed the projected life of the structure. If this is not the case, however, then Stages 2 and 3 can occur. In these circumstances, comparing the durability of plain and galvanized steel, results from the present study would indicate that their performance would be similar provided protection from the alkalinity of the concrete remained and, therefore, Stage 1 would be identical for both types of reinforcement. Variations in time-to-cracking would, therefore, result from differences in the rates in Stages 2 and 3. It seems probable that Stage 3 is similar for both reinforcement materials, as the present study indicated that cracking of concrete following corrosion of galvanized steel is subsequent to corrosion of the substrate steel; therefore, Stage 2 would be the rate-controlling step. In this circumstance the time to first maintenance would be controlled by the corrosion resistance provided by the reinforcement during Stage 2 which, in the case of galvanzied steel, would be related to the corrosion resistance of the zinc coating. The greatest corrosion protection conferred by zinc is in environments lying within the pH range 8 to 12 [*16*], which is the pH range normally found in alkaline and carbonated concrete.

It would seem, therefore, that the use of galvanized coatings to provide additional protection to steel in concrete gains the most benefit under conditions where the rate of corrosion of zinc is at its minimum. These conditions are most likely to exist when the pH range of the concrete falls between 8 and 12, and are most likely encountered in good-quality uncarbonated concrete, or permeable but chloride-free concrete where carbonation has reached the level of the reinforcement. There also appears to be some benefit in the use of galvanized reinforcement in concrete containing chlorides in low or moderate concentration (up to 0.96 percent chloride ion) but where the level of chloride ion lies at about 1.9 percent or above, although some amelioration of the corrosion is apparent, results suggest that only a short-term benefit is achieved.

Conclusions

1. At high chloride levels (1.9 percent and above) serious corrosion of both galvanized and plain steel occurs. Galvanizing provides some delay in the onset of corrosion-induced cracking of the cover, but this is of only limited benefit, the delay ranging between six months and one year.
2. In chloride-bearing concrete containing up to 0.96 percent chloride, galvanizing provides a more tangible benefit, but the full extent of this cannot be quantified within the present phase of this program.
3. In concrete to which no chloride has been deliberately added, but which is subject to extensive carbonation (as in lightweight-aggregate concrete), the application of a galvanized coating to the steel is of benefit in measurably delaying the onset of cracking.
4. In good-quality concrete made without chloride, bright, shotblasted, and galvanized steel all perform well. There is little corrosion on the steel, and in the case of zinc the loss is small. Where prerusted steel is used in similar concrete, corrosion rates are higher, but at the present stage of testing no cracking has resulted.
5. The corrosion of steel in concrete that leads to cracking of the cover can be characterized by three stages: initiation, corrosion, and cracking of the cover as a result of development of tensile forces in the cover produced by the formation of expansive corrosion products. Where the initiation phase of the process is shorter than the predicted life of the structure, the resultant corrosion can have a significant effect on maintenance requirements. In this circumstance the corrosion resistance of the reinforcement becomes of major importance. Results from the present study indicate that under certain circumstances the use of galvanized reinforcement can considerably lengthen the second phase of the corrosion process, thus providing a useful benefit.

Acknowledgments

The authors would like to note the work of Dr. L. H. Everett and the late Mr. R. D. J. Tarleton in setting up the Series I exposure program.

The work described has been carried out as part of the research program of the Building Research Establishment of the Department of the Environment and this paper is published by permission of the Director.

APPENDIX I

Weight Loss and Mean Thickness Calculations

Corrosion of the reinforcement was measured initially by weight loss. Both Series I and II prism specimens were exposed containing preweighed reinforcement which allowed weight loss to be measured following removal of specimens from exposure and cleaning of test bars at the end of the exposure period.

The basic calculation for percentage weight loss was

$$\%\text{ Weight loss of reinforcement} = \left[\frac{\text{Preexposure weight} - \text{post-exposure weight (after cleaning)}}{\text{Preexposure weight}}\right] \times 100 \quad (1)$$

For Series I prisms containing:

1. Bright mild steel: the calculation is as Eq 1.
2. Prerusted mild steel: the preexposure weight used in Eq 1 was obtained from

$$S - \frac{2}{3}(R - S)$$

where S is the initial bright weight of bar and R is the rusty preembedment weight. To apply this calculation, all the bars were adjusted to an initial common preexposure weight (according to bar size). The bars then were subjected to several months' exposure at a rural site to induce the preembedment rust. After this exposure they were reweighed prior to embedment in the prisms. In the calculation it has been assumed that the rust formed during the preembedment exposure contained 67 percent iron.

3. Hot-dipped galvanized mild steels: weights were measured before and after galvanizing, enabling individual coating weights to be calculated. As none of the bars from the Series I prisms showed serious zinc loss, it was possible to employ the post-galvanized weight for preexposure weight in Eq 1. The postexposure weight was measured on galvanized bars after exposure.

For Series II prisms containing:

1. Shotblasted high-yield steel: the calculation is as in Eq 1.
2. Prerusted high-yield steel: individual amounts of rust on each bar were unknown. However, using preexposure weights for shotblasted bars and the prerusted weights, the mean proportion of rust on the rusted bars was estimated to be 0.518 percent by weight. The preexposure weight used in Eq 1 was obtained from (0.9948 × preexposure rusty weight) and assumed uniform rust on all bars.
3. Galvanized high-yield steel: the pregalvanized rod weights were unknown, so a correcting factor assuming uniform zinc coating was developed from which percentage zinc loss could be determined. However, provided some zinc remained on the test bar and the exposure weight loss was converted to thickness loss, such correction was unnecessary and comparisons of steel and zinc loss could be made.

The exposure weight loss calculated from Eq 1 was converted to volume of reinforcement loss and, from thence, to thickness loss using relative densities of the appropriate metal and the mensuration surface area of the bar. Series I bars were simple cylinders; Series II bars were deformed cylinders, the deformations adding 15 percent area to that of the simple cylinder. Thickness loss was calculated using the following formula

$$\text{Mean thickness loss} = \frac{\text{Mass loss on exposure (g)} \times 10^4}{\text{Relative density} \times \text{surface area of test bar (cm}^2\text{)}}$$

The relative density for zinc was taken as 7.14g cm^{-3} and for steel as 7.86 g cm^{-3}. The surface areas of the test bars were as follows:

	Cover	Length	Area, cm^2	Ungalvanized Weight, g
Series I	2 in.	8 in.	83.6	198.00
	1½ in.	9 in.	93.74	223.00
	1 in.	10 in.	103.87	248.20
	½ in.	11 in.	114.01	273.50
Series II	20 mm	260 mm	91.94	...
	10 mm	280 mm	99.02	...

APPENDIX II

Assessment of Advantage of Chromate Pretreatment of Galvanized Reinforcement

The action of a chromate pretreatment is to supplement any action of inhibition from soluble chromic oxide present in the cement, suppressing hydrogen evolution on the zinc surface [5] with a possible increase in durability. Half the galvanized bars used in the Series II specimens were chromated before embedding. The procedure consisted of a 10 to 20 s dip in acid sodium dichromate solution (consisting of 150-g sodium dichromate dissolved in 10-g concentrated sulphuric acid and diluted to one litre with distilled water) followed by air drying.

The effectiveness of the treatment is difficult to show in the short term in uncontaminated good-quality concrete, as zinc performs well under these conditions. However, a comparison of all the thickness loss data relating to the chromated and unchromated bars exposed under equivalent conditions demonstrates that chromating can reduce zinc loss. Statistical analysis using linear regression gave good correlation, the relationship being of the form:

$P = 2.97 + 1.13\ C$
P = zinc loss for untreated bars
C = zinc loss for chromated bars

From this relationship equivalent losses can be predicted for equivalent conditions, the calculation producing the following data

$P_{(\mu m)}$	$C_{(\mu m)}$
4	1
14	10
116	100

This holds for the conditions of test, that is, Ordinary Portland Cement with 0.002 percent soluble chromate. Chromating is likely to be most advantageous in cements with low soluble chromate contents.

TABLE 8—*Mix ratios used.*

Bulk Density Used, kg/m^3	Cement OPC[a] Parts by Weight — 1440	Concreting Sand < 3/16-in. Parts by Weight — 1600	Flint-Gravel 3/16 to 3/4 in. Parts by Weight — 1440	Cement Content, kg/m^3	Bags of Cement per Cubic Yard	W/C[b] Ratio	Gallons/Bag
Series I	1	1.5	3	270	4.8	0.5	4.7
	1	2	4	210	3.8	0.6	5.6
	1	3.5	4.5	170	3.0	0.9	8.5
Series II	1	2.4	3.6	220	4.0	0.6	5.6
	1	3.2	4.8	170	3.0	0.75	7.0
	1	2	3	250	4.4	0.55	5.2

[a] OPC = Ordinary Portland Cement.
[b] W/C = water/cement.

APPENDIX III

Conversion of U.K. to U.S. Units

Definitions

Cement:Aggregate Ratio—The ratio of the mass of cement to the mass of total aggregate (that is, sand plus coarse aggregate) in a concrete mix.

Cement Content—The mass in kilograms of cement contained in 1 m^3 of fresh, fully compacted concrete.

Water/Cement (W/C) Ratio—The ratio of the mass of total water (that is, mix water plus free water content of aggregates) to the mass of cement in a concrete mix.

The mix ratios and chloride additions used are given in Tables 8 and 9.

TABLE 9—*Chloride additions used.*

			Chloride Additions (% $CaCl_2$ Anhydrous by Weight Cement)			
			½%	1½%	3%	5%
Mix Ratio		Bags/yd^3	lbs Cl^- ion per yd^3			
	1:8	3	0.9	2.7	5.4	8.9
Series II	1:6	4	1.2	3.6	7.3	12.1
	1:5	4.4	1.3	4.0	...	13.4

NOTE: The standard U.S. bag of cement is taken as 94 lbs.

References

[1] *Zinc-Coated Reinforcement for Concrete*, BRE Digest 109 Building Research Establishment, U.K., 1969.
[2] Porter, F. C., *Concrete*, Vol. 8, 1976, p. 29.
[3] Stark, D. "Galvanised Reinforcement in Concrete Containing Chloride," Portland Cement Association International Lead Zinc Research Organization, Report on Project ZE-247, July 1977.
[4] Bird, C. E., *Corrosion Prevention and Control*, Vol. 11, No. 7, 1964, pp. 17-21.
[5] Everett, L. H. and Treadaway, K. W. J., "The Use of Galvanized Steel Reinforcement in Building," Building Research Station Current Paper CP3/70, Garston, Herts, 1970.
[6] Bird, C. E. and Strauss, J. F., *Materials Protection*, Vol. 6, No. 7, 1967, pp. 48-52.
[7] Boyd, W. K. and Tripler, A. B., *Materials Protection*, Vol. 7, No. 10, 1968, pp. 40-47.
[8] Duval, R. and Arliguie, G., *Memories Scientifique de la Revue de Metallurgie*, Vol. 71, No. 11, 1974, pp. 719-727.
[9] Unz, M., *Journal of the American Concrete Institute*, Vol. 75 March 1978, pp. 91-98.
[10] Griffen, D. F., "The Effectiveness of Zinc Coating on Reinforcing Steel in Concrete Exposed to a Marine Environment, U.S. Naval Civil Engineering Laboratory Technical Note N-1032, July 1969.
[11] Baker, E. A., Money, K. L., and Sanborn, C. B. in *Chloride Corrosion of Steel in Concrete, ASTM STP 629*, American Society for Testing and Materials, 1976, pp. 30-50.
[12] Okamura, H. and Hisamatsu, Y., *Materials Performance*, Vol. 15, No. 7, 1976, pp. 43-47.

[*13*] Bird, C. E. and Callaghan, B. D. in *Proceedings,* Conference on Concrete in Aggressive Environments, South African Corrosion Institute, Oct. 1977.
[*14*] Cooke, A. R. and Radtke, S. F. in *Chloride Corrosion of Steel in Concrete, ASTM STP 629,* American Society for Testing and Materials, 1976, pp. 51-60.
[*15*] Beeby, A. W., "Cracking and Corrosion, Concrete in the Oceans," Technical Paper No. 1, Cement and Concrete Association, Wexham Springs, U.K., 1978.
[*16*] Rotheli, B. E., Cox, G. L., and Littreal, W. B., *Metals and Alloys,* Vol. 3, 1932, pp. 73-76.

D. Stark[1]

Measurement Techniques and Evaluation of Galvanized Reinforcing Steel in Concrete Structures in Bermuda

REFERENCE: Stark, D., "**Measurement Techniques and Evaluation of Galvanized Reinforcing Steel in Concrete Structures in Bermuda,**" *Corrosion of Reinforcing Steel in Concrete ASTM STP 713,* D. E. Tonini and J. M. Gaidis, Eds., American Society for Testing and Materials, 1980, pp. 132–141.

ABSTRACT: The performance of galvanized reinforcing steel in concrete structures exposed to seawater in Bermuda was evaluated by measurements of chloride concentrations in the concrete, and average depths of corrosion of the galvanized coatings. Results indicate that little more than superficial corrosion of the coatings has occurred in 7- to 23-year-old normal-quality concretes containing as much as 10 times the chloride concentrations needed to induce corrosion of untreated steel. In all but one case the outer free zinc layer was still present on the coating. In these instances, the average depths of corrosion ranged from zero to 0.013 mm, with the amount of coating remaining ranging from 92 to 100 percent of the original thickness. Localized corrosion to the steel substrate was found only in uncompacted highly porous concrete in a poorly bonded cold joint.

KEY WORDS: corrosion, galvanized, reinforcing steel, concrete, chloride, metallographic

Research during the past 10 to 20 years has revealed the significance of chloride ion in the corrosion of untreated reinforcing steel in concrete [*1,2*].[2,3] These studies have shown that, where the chloride ion concentration is above a certain threshold level, corrosion may develop in the presence of

[1]Principal research petrographer, Concrete Materials Research Department, Portland Cement Association, Skokie, Ill. 60077.

[2]Lewis, D. A. in *Proceedings,* First International Congress on Metallic Corrosion, 1962, pp. 547–555.

[3]Clear, K. C. and Hay, R. E., "Time-To-Corrosion of Reinforcing Steel in Concrete Slabs," Report No. FHWA-RD-73-32, Federal Highway Administration, Washington, D. C., April 1973.

moisture and available oxygen. This threshold level may be as low as about 0.20 percent total chloride by weight of cement, or roughly 0.025 percent by weight of normal structural-quality concrete. Numerous efforts also have been made to develop measures for preventing corrosion in new structures, and rehabilitating already deteriorated structures to prevent further corrosion problems. Among the recommended measures are the use of epoxy-coated reinforcing bars, the application of cathodic protection systems, the use of lower-permeability concrete, increased concrete cover over the steel, and galvanized coatings on the reinforcing steel. Of these measures, the use of galvanized reinforcing steel has one of the more extensive performance records in concrete structures. Much of this service history has been developed in Bermuda, where galvanized steel has been used since the early 1950's and has been prescribed for many years as a matter of policy by the Bermuda Department of Public Works.

The purpose of the investigation described herein was to identify certain pertinent conditions in concrete structures in Bermuda where galvanized reinforcing steel was used, and to determine the corrosion resistance of the galvanized coating under these conditions.

Selection of Structures

Two factors were of primary consideration in the selection of structures for study: age, and severity of exposure, based on concentration of chloride ion. Review of available structures suggested that those in direct contact with seawater should provide the most favorable exposures for study. The following four structures, ranging in age from 7 to 23 years, were chosen for study.

1. Longbird Bridge abutment. This structure was built in 1953. Specimen locations are on the west face of the south abutment, 90 to 150 cm above the high-tide level. The concrete contains No. 6 round deformed bars at a depth of 11.5 cm from the vertical face of the abutment.
2. St. George Dock, St. George. This structure was built in three stages, in 1964, 1966, and 1969. Specimen locations are on the vertical face of the seawall, at mean tide, to 30 cm above the high-tide level. Number 5 round deformed bars are located at depths of 5.7 to 7.0 cm from the vertical face.
3. Hamilton Dock, Hamilton. This dock was built in 1966. Specimen locations are on the vertical face of the seawall, ranging from 45 cm below to 30 cm above the high-tide level. Number 8 twisted square bars are located at depths of 5.7 to 7.0 cm from the vertical face.
4. Royal Bermuda Yacht Club jetty, Hamilton. This jetty was built in 1968. Specimen locations are 30 cm below high tide on the north face at the east end of the jetty. Number 3 smooth round bars are located at a depth of 6.3 cm from the vertical face.

Sampling and Testing Procedures

The relationship between chloride concentrations in the concrete and severity of corrosion of the galvanized coating on the reinforcing steel was considered the most relevant parameter for evaluating the corrosion resistance of the galvanized reinforcing steel. In addition, petrographic examinations were deemed necessary to disclose any evidence of corrosion-related distress in the concrete. For this work, 10-cm-diameter concrete cores were procured from the structures and locations noted in the preceding. Most cores were taken sufficiently deep to retrieve a section of reinforcing steel but, in several cases, an adjacent core was taken between reinforcing bars to a depth greater than the steel to provide for an intact concrete specimen for chloride analyses at these greater depths. Because seawater had to be used for coring, it was anticipated that additional chloride would permeate the concrete cores and distort the true chloride concentrations of the concrete in place. However, preliminary analyses revealed that chloride contamination from this source extended inward from the cylindrical core surface only to a depth of 0.6 to 1.0 cm. Therefore, specimens for chloride analyses were obtained from the cores from well within this outer cylindrical zone.

Powder specimens for chloride analyses were procured from the cores after shipment to the Portland Cement Association (PCA) laboratories. Here, the cores were positioned in a lathe, and 1.0- or 3.2-cm-diameter carbide-tip drills were used to drill out the powder specimens at preselected 0.6-cm intervals. Two or three companion specimens were taken from the same core where the smaller drill was used. All sampling was done without the aid of a liquid coolant.

Chloride measurements were made on the powder specimens after final screening through a No. 100 sieve. A 10-g specimen was then weighed, digested in dilute nitric acid at incipient boiling for 15 min, filtered, and the filtrate titrated potentiometrically with 0.1*N* silver nitrate solution. Nonevaporable water contents were measured to correct for nonuniformities in paste-aggregate ratios among specimens from a given structure.

The depth of corrosion of the galvanized coating on the reinforcing steel was determined from measurements utilizing a metallographic microscope. Short lengths of reinforcing steel, taken during coring, were sectioned transversely, finely polished, and etched to reveal the steel substrate and individual layers in the galvanized coating. Figure 1 illustrates a typical polished surface of galvanized reinforcing steel. In most cases, a section of reinforcing bar was selected which still carried an adhering layer of mortar, so as to provide a more definitive base for measuring the depth of corrosion of the coatings. To obtain an average coating thickness, microscopic measurements were made on 10 random areas on each specimen, using a magnification of 16 to 40 times, depending on coating thickness.

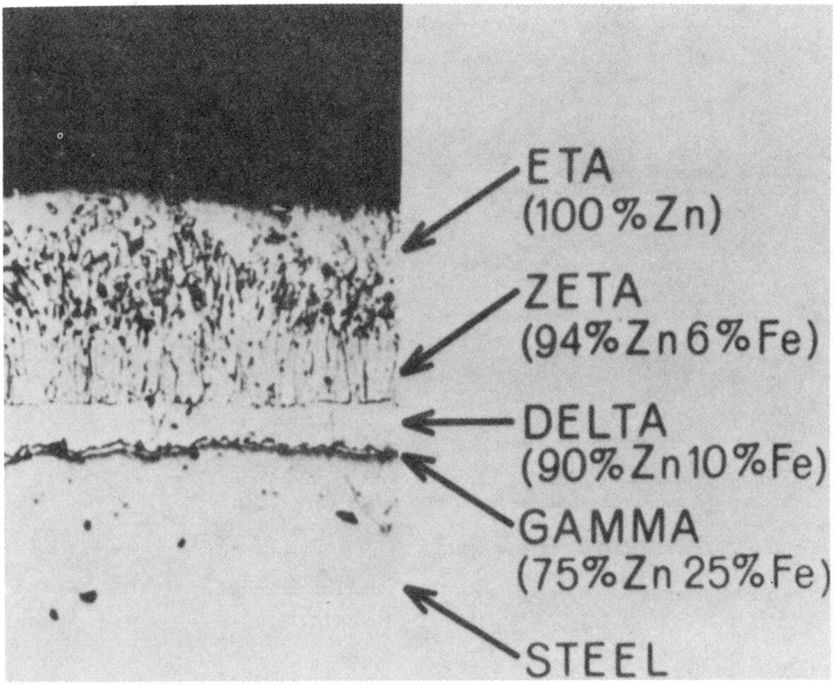

FIG. 1—*Illustration of a typical uncorroded galvanized coating on reinforcing steel.*

The concrete cores from each location were examined petrographically for evidence of corrosion reaction products and associated microcracking that might have been induced by corrosion.

Results of the Investigation

Figures 2-4 show the depth, from the exposed surface, of the galvanized reinforcing bars with respect to chloride concentration gradients in the concrete. Also shown, by the dashed line, is the threshold level generally considered necessary to induce corrosion of untreated reinforcing steel. In all cases, the steel bars were in concretes with chloride concentrations well above the threshold level of 0.025 percent, which is generally considered necessary to induce corrosion of untreated steel. Assuming that only 75 percent of these values represent water-soluble chloride, or that fraction available for the corrosion process, the chloride concentrations are still well above the corrosion threshold. Thus, all locations selected for sampling provide exposure conditions conducive to the development of corrosion of untreated steel.

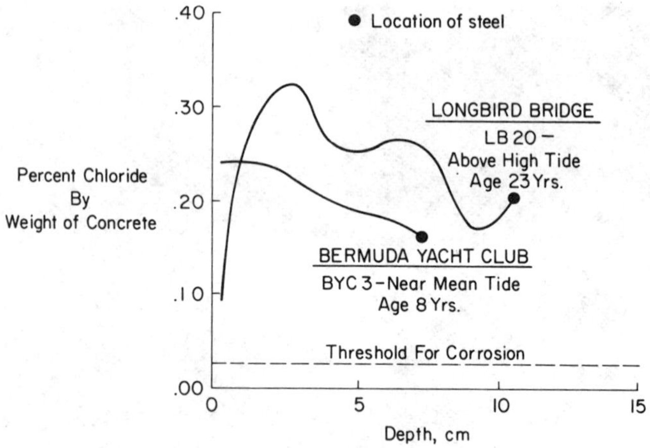

FIG. 2—*Chloride concentrations in concrete adjacent to galvanized reinforcing steel in the Longbird Bridge abutment and the Royal Bermuda Yacht Club jetty.*

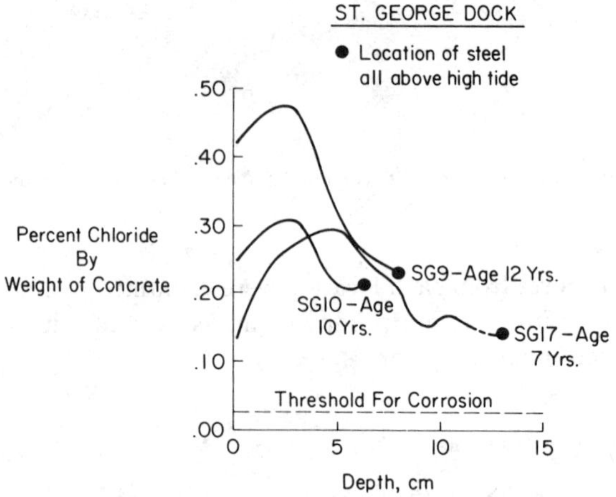

FIG. 3—*Chloride concentrations in concrete adjacent to galvanized reinforcing steel in St. George Dock.*

A summary of data from the chloride and metallographic measurements is given in Table 1. Here, comparisons of chloride concentrations in concrete adjacent to steel with average thickness of zinc corrosion layer indicate that little or no corrosion has occurred, even in concrete more than 20 years old. Assuming available chloride levels are 75 percent of the total chloride content, chloride concentrations in the Longbird Bridge abutment are about six times as great as the threshold necessary to induce corrosion of untreated steel, but the average zinc corrosion layer thickness has reached only 0.005

mm, thus leaving at least 98 percent of the original coating unaffected and intact. The greatest extent of corrosion has occurred in 12-year-old concrete from the St. George Dock where, with eight times the threshold level of chloride ion present, an average corrosion layer thickness of 0.013 mm has developed. Here the remaining coating thickness averages about 92 percent of the original thickness. In the younger concretes from the St. George Dock, lower chloride concentrations are present and lesser thicknesses of corrosion layer have developed. In these instances, 98 to 99 percent of the original thicknesses remain. Data for the 10-year-old Hamilton Dock concrete reveal chloride concentrations of two to more than four times that necessary for corrosion of untreated steel, but development of corrosion layers only 0.005 to 0.008 mm thick. Here the more severe exposure occurs at a level slightly

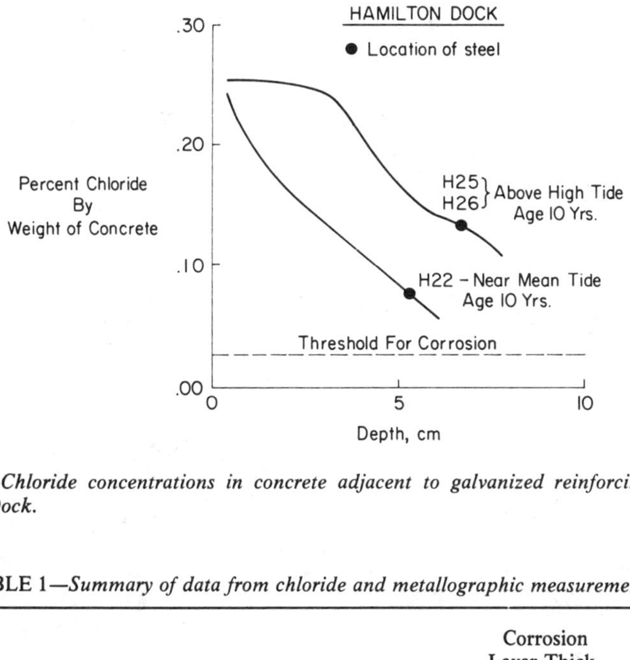

FIG. 4—*Chloride concentrations in concrete adjacent to galvanized reinforcing steel in Hamilton Dock.*

TABLE 1—*Summary of data from chloride and metallographic measurements.*

Structure	Age, years	Core No.	% Cl^- by Weight	Corrosion Layer Thickness Avg, mm	% of Coating Remaining
Longbird Bridge	23	LB20	0.19	0.005	98
St. George Dock	12	SG9	0.27	0.013	92
	10	SG10	0.22	0.005	99
	7	SG17	0.14	0.002	98+
Hamilton Dock	10	H22	0.08	0.005	95
	10	H26	0.14	0.008	96
Bermuda Yacht Club	8	BCY3	0.16	none	100

above high tide, compared with mean tide, and upward of 96 percent of the original average thickness of the galvanized coating is still present on the steel bar. In the eight-year-old concrete at the Royal Bermuda Yacht Club, chloride concentrations have reached nearly five times the threshold, but essentially no corrosion has occurred. Although chloride levels have reached as much as eight times the level necessary to corrode untreated steel, there has been little more than superficial corrosion of the galvanized coatings on the reinforcing steel.

Photomicrographs of sections of the various galvanized reinforcing bars illustrate the high corrosion resistance of the galvanized coating in these seawater exposures. Figure 5 shows a section of reinforcing steel in Core LB20, taken from the Longbird Bridge abutment. After 23 years, with more than six times the chloride concentration needed to induce corrosion of untreated steel, the (outer) free zinc layer is still present, and the underlying zeta and delta layers of the galvanized coating are still intact. Figure 6 illustrates the galvanized coating on the reinforcing bar in Core SG9 from St. George Dock, where the greatest thickness of corrosion layer has developed. Here the free zinc (eta) layer has corroded away, leaving the zeta and delta layers directly exposed to the chloride solutions in the concrete. Figure 7 is an illustration of the galvanized coating on the reinforcing bar in Core BYC3,

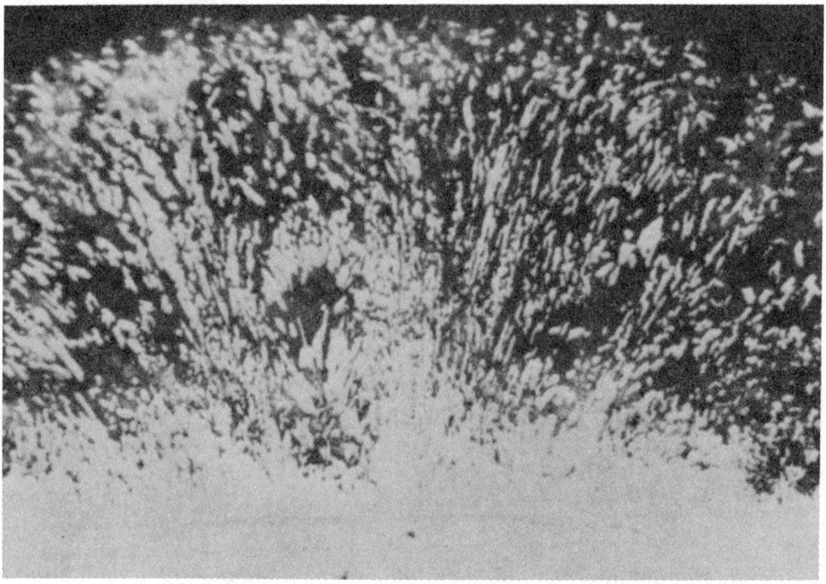

FIG. 5—*Photomicrograph of galvanized coating on reinforcing bar in Core LB20 from the Longbird Bridge abutment* (×*250*).

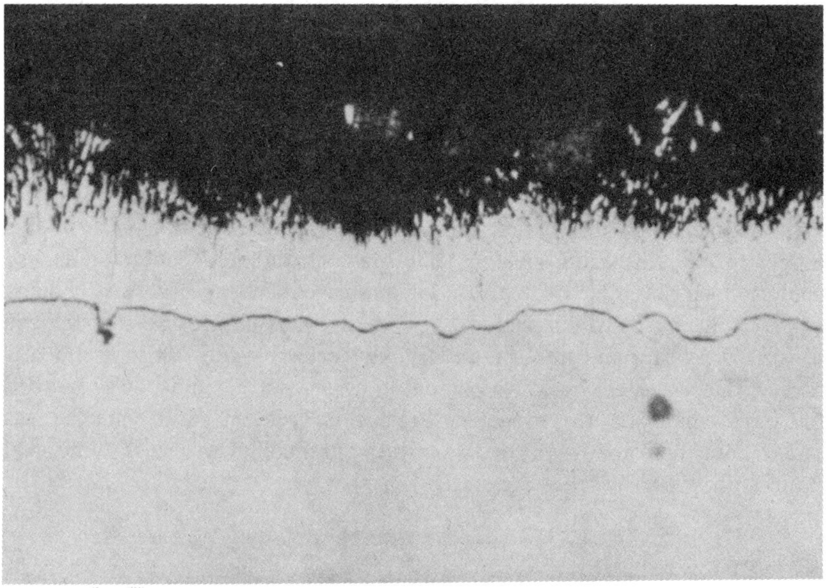

FIG. 6—*Photomicrograph of galvanized coating on reinforcing bar in Core SG9 from St. George Dock* (×*240*).

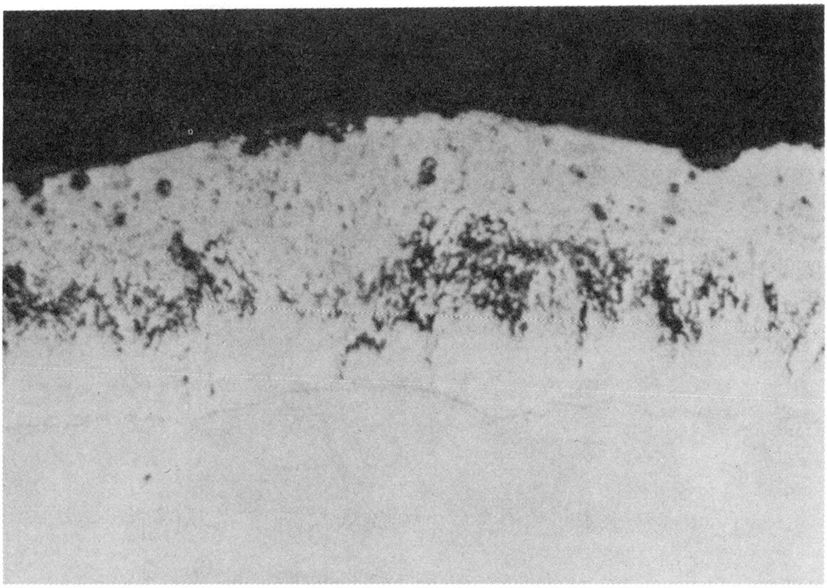

FIG. 7—*Photomicrograph of galvanized coating on reinforcing bar in Core BYC3 from the Royal Bermuda Yacht Club jetty* (×*360*).

taken from the Royal Bermuda Yacht Club jetty. Here, there is no corrosion of the coating.

Petrographic examinations were made of the concrete in each core to determine also if there is distress associated with corrosion of the galvanized reinforcing steel. These examinations failed to reveal any such evidence; thus, the minor corrosion that has occurred has not been detrimental to the quality of the concrete.

Although the foregoing data confirm the highly satisfactory performance of galvanized reinforcing steel in chloride-contaminated concrete, its use should not be taken as a substitute for quality concreting practices. This is indicated in Fig. 8, which shows that corrosion has progressed through the galvanized coating and into the underlying steel substrate. This specimen is from St. George Dock, and was taken on a very porous and poorly bonded cold joint where an uncompacted, high water/cement ratio concrete was used. Under these conditions, seawater had practically free access to the steel with resulting corrosion of the steel.

Summary

The combined results of chloride concentration measurements in the concrete and measurement of average depths of corrosion indicate that galva-

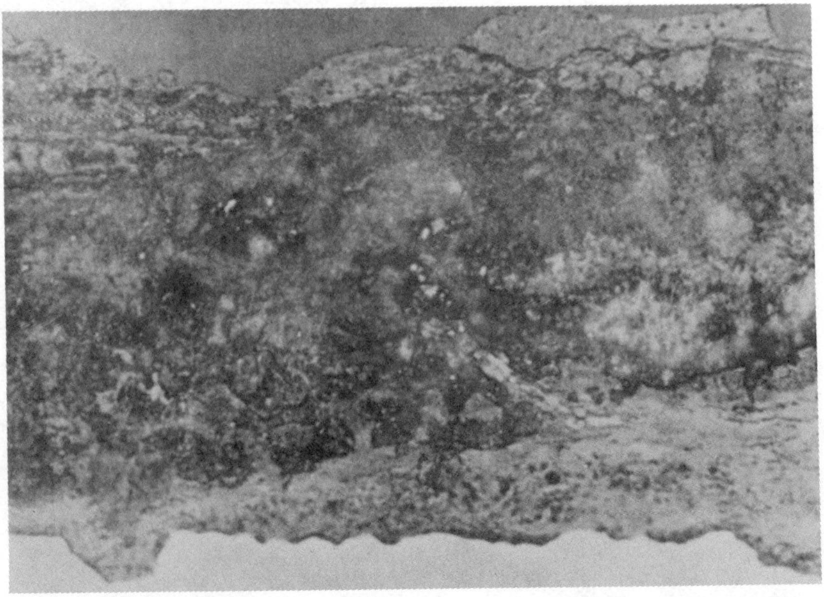

FIG. 8—*Photomicrograph of reinforcing bar from the St. George Dock concrete where corrosion has progressed to the steel substrate at a porous cold joint in the seawall* (×*220*).

nized reinforcing steel in concrete in Bermuda has shown high degrees of corrosion resistance under conditions where untreated steel would have been expected to display severe corrosion. At locations where normal-quality concrete was sampled, averages of 92 to 100 percent of the original coating thicknesses are present after 7 to 23 years' exposure of the concrete to seawater. The only evidence of steel corrosion in concrete was at a cold joint where uncompacted, high water/cement ratio concrete was present. In all cases, however, there was no corrosion-associated distress in the concrete.

Acknowledgments

The author wishes to acknowledge the staff of St. Joe Minerals, who performed the metallographic measurements on the galvanized coatings on the reinforcing steel.

D. J. H. Corderoy[1] *and H. Herzog*[2]

Passivation of Galvanized Reinforcement by Inhibitor Anions

REFERENCE: Corderoy, D. J. H. and Herzog, H., **"Passivation of Galvanized Reinforcement by Inhibitor Anions,"** *Corrosion of Reinforcing Steel in Concrete, ASTM STP 713,* D. E. Tonini and J. M. Gaidis, Eds., American Society for Testing and Materials, 1980, pp. 142–159.

ABSTRACT: Concrete reinforced by galvanized reinforcing bars requires the addition of inhibitor anions to passivate the zinc; otherwise evolution of hydrogen reduces the bond strength. In the present study, zinc in saturated calcium hydroxide solution, which is the principal electrolyte formed on initial hydration of cement, was found to passivate in the presence of either sodium chromate or chromic oxide. In both cases chromate ions were reduced instead of water so that there was no evolution of hydrogen. The passive film consisted of zinc chromate and chromic oxide, but calcium hydroxo-zincate was also produced, which assisted in passivating the zinc surface. In the case of chromic oxide a smaller concentration of chromate ions was present and passivated the zinc to a lesser extent. While 70 ppm of sodium chromate was sufficient to passivate the zinc, at least 300 ppm of chromic oxide was necessary to achieve the same degree of passivation. Chloride ions in the presence of chromate ions compete for the zinc surface and a critical concentration of chromate is necessary. This observation has been correlated with the passivation of galvanized reinforcing bars in concrete and it is considered that passivation is best achieved by the addition of 70-ppm sodium chromate to the concrete mix.

KEY WORDS: corrosion, concrete, mortar, reinforcing steels, galvanized reinforcement, chlorides, passivation, inhibitor anions

While the cover provided by concrete is normally sufficient to protect steel reinforcement from corrosion, there are a number of situations, such as exposure to a chloride environment or where sufficient cover cannot be provided such as in architectural fascia, when it is necessary to protect the steel reinforcement, and the most common method is galvanizing either in the form of hot-dip galvanizing or zinc-rich paint. Such zinc coatings have a number of advantages. For one, zinc provides sacrificial protection

[1]Senior lecturer, School of Metallurgy, The University of New South Wales, Kensington, Australia, 2033.

[2]Materials engineer, Technisearch Limited, Collilngwood, Australia, 3066.

of the steel, and, additionally, zinc can withstand higher chloride ion concentrations than steel and can also withstand lower pH levels in the concrete resulting from carbonation. Moreover, the corrosion products of zinc occupy less volume than those of steel so that for equivalent cover the life of concrete reinforced with galvanized steel is increased by at least some 50 percent [*1*].[3]

Zinc being anodic to steel cathodically protects steel, the anode half-cell reaction being

$$Zn \rightarrow Zn^{2+} + 2e^-$$

with the zinc ions reacting with hydroxyl ions to form zinc hydroxide

$$Zn^{2+} + 2OH^- \rightarrow Zn(OH)_2$$

At the steel reinforcement the cathode reaction and the formation of hydrogen gas occurs in accordance with

$$2H^+ + 2e^- \rightarrow H_2 \text{ (gas)}$$

Thus, while the steel is cathodically protected by the zinc coating, the hydrogen gas so liberated envelopes the bar. This reaction occurs most readily when the galvanized bars are first cast in the wet concrete so that a spongy or foam-like concrete surrounds the bar. However, such evolution of hydrogen can be eliminated by passivation of the zinc with chromate ions.

Passivation of the zinc also occurs by the formation of calcium hydroxo-zincate on the surface [*2*] when the zinc is exposed to calcium hydroxide in solution or in wet concrete. Rehm et al [*3*] observed that the rate of the formation of calcium hydroxo-zincate on metallic zinc is quite high but is accompanied by the evolution of hydrogen, which reduces the bond strength. Bird [*4*] also observed that hydrogen was liberated when galvanized steel was embedded in concrete, which in turn became spongy in the immediate volume of surrounding concrete so that there was a loss of bond strength between the concrete and the reinforcement. Bird [*5*], however, established that the hydrogen evolution could be inhibited by the small amounts of chromate naturally present in some cements. The evolution of hydrogen could, moreover, be prevented by rinsing the galvanized rods with a 20 percent chromic acid solution or with a 6-ppm chromic oxide in a sodium hydroxide solution having a pH of 12.7, the evolution of hydrogen being basically eliminated by the preferential reduction of chromate rather than water. At pH values of approximately 12.5, chromium

[3]The italic numbers in brackets refer to the list of references appended to this paper.

trioxide (CrO_3) is reduced to a chromic oxide (Cr_2O_3) with the formation of a passive film consisting of zinc chromate, chromic oxide, and possibly zinc hydroxide or calcium hydroxo-zincate. However, the exact nature of the passivation of zinc by chromates has not been established in an alkaline environment.

There is, furthermore, some disagreement as to the quantity of chromium or chromate compounds required for the elimination of hydrogen evolution and total passivation of the zinc. In 1974 an International Lead Zinc Research Organization (ILZRO) publication [*6*] recommended the addition of 300 ppm by weight of chromium trioxide to the mix water to passivate the zinc, while in 1976 Cook [*7*] recommended the addition of 10^{-3} M potassium chromate. On the other hand, Mahaffey [*8*], Bracket et al [*9*], and Bird [*5*] report that 70 ppm of chromium trioxide is required. One of the intentions of the present work was the clarification of the quantity and efficacy of these compounds required to inhibit the formation of hydrogen gas.

When chlorides are present, however, the levels of chromate used to provide inhibition for hydrogen evolution from galvanized reinforcement may not be sufficient and a double log/log relationship exists for concentrations of aggressor and inhibitor anions. Nevertheless, even in the absence of chromate ions, zinc remains passive in chloride solutions of much higher concentrations than those eliminating the passivity of steel [*10*]. For instance, Cornet et al [*11*] observed that the passivity of black steel broke down at sodium chloride concentrations of approximately 0.2 weight percent, while galvanized steel was able to withstand concentrations of 2 weight percent but once passivity was lost the zinc corroded in preference to steel.

Experimental

The aim of the present experiments was to determine the extent to which galvanized reinforcing bars are passivated in saturated calcium hydroxide solution. The corrosion behavior of reinforced concrete can be simulated to some extent in such a solution, as the electrolyte in concrete has been shown to consist of saturated calcium hydroxide as well as sodium and potassium hydroxides buffered at a pH of 12.5. Furthermore, it was intended to study the effect of inhibitor anions on the passivity and to determine the efficacy of such inhibitors in the presence of aggressor chloride ions. Two types of experiments were conducted: first, accelerated corrosion tests of both black and galvanized reinforcing bars encased in test cylinders of cement mortar with or without inhibitor additions; and second, the passivation and polarization of zinc in saturated calcium hydroxide solution were studied using a range of inhibitor additions together with a range of sodium chloride concentrations.

Accelerated Corrosion Tests

Cylindrical cement mortar test specimens were prepared using a mix having a cement factor of 323 kg/m^3 and a water cement ratio of 0.40. Analysis of the cement showed a natural chromate content of 5 ppm. Three batches of mortar were prepared containing, respectively, 70-ppm sodium chromate, 70-ppm chromic oxide, while the third batch was free of inhibitor additions. The cement mortar test cylinders were compacted by mechanical vibration and cured in water for a period of 14 days prior to the commencement of testing.

The cement mortar test cylinders were cast into molds 200 mm in length using 16-mm deformed cold-twisted grade 410C and hot-rolled deformed structural-grade 230S bars to Australian Standard 1302-1977, Steel Reinforcing Bars for Concrete. The bars were in both the black and hot-dip galvanized condition and had a uniform depth of cover of 10 mm.

The test cylinders were subjected to accelerated corrosion tests in a 3 percent sodium chloride solution with an impressed constant current of 3.0 mA, which represented a current density of 0.36 amp/m^2.

Passivation and Polarization of Zinc in Saturated Calcium Hydroxide Solution

In these experiments potential was measured with respect to time while separate galvanostatic polarization experiments were also conducted. A number of working zinc electrodes 1 cm^2 was mounted in cold-setting resin so that only one face would be exposed to the electrolyte. The electrolyte was a saturated calcium hydroxide solution of pH 12.5 containing 10^{-3} M of potassium sulphate. In the various experiments sodium chromate, chromic oxide, and sodium chloride were added to the calcium hydroxide solution.

In the measurements of zinc potential with respect to time, a saturated calomel electrode was used as a reference and all measured potentials were converted relative to a standard hydrogen electrode. Initially, measurements were made with the solutions containing 70-ppm and 300-ppm sodium chromate, respectively. The experiments were continued with additions of sodium chloride being added to give chloride ion concentrations of 10^{-3} M, 10^{-2} M, 10^{-1} M and $10^{-0.5}$ M. Subsequently, the experiments were repeated using 70 and 300 ppm of chromic oxide and the same concentrations of chloride ion. While all these experiments were performed on clean zinc electrodes, the effect of previously passivated zinc electrodes was also examined. In the former case, the electrodes were polished so that clean equipotential surfaces were obtained, but in the latter case the same electrodes were allowed to passivate for 5 h in their respective solutions of calcium hydroxide with either 70-ppm sodium chromate or 70-ppm chromic oxide. For all these experiments the current through the ammeter was maintained at zero.

Galvanostatic polarization experiments were made using the same circuit but with a platinum counter electrode and a variable resistor to control the current. Previously passivated zinc electrodes were used in saturated solutions of calcium hydroxide with additions of either 70-ppm sodium chromate or 70-ppm chromic oxide and with subsequent additions of sodium chloride to give similar chloride concentrations to those of the previous experiments.

Finally, infrared spectroscopic analyses were made on films mechanically removed from the zinc, which had been immersed in saturated calcium hydroxide containing either sodium chromate of chromic oxide. Similar infrared spectroscopic analyses were made on films mechanically removed from reinforcing bars subject to accelerated corrosion tests.

Results

Corrosion Behavior of Reinforced Cement Mortar Test Cylinders

The effect of corrosion on the reinforcing bars was determined by the time taken for rust stains from the corroding bars to appear on the mortar surface, and also by the appearance of the reinforcing bar and the cement mortar-reinforcing bar interface after cutting the cylinders open. Under the high impressed current density of 0.36 amp/m^2 the sequence of failure in terms of the first appearance of red rust staining on the outside of the cement mortar cylinders was as given in Table 1.

Inspection of the cement mortar reinforcing bar interface after cutting with a diamond saw indicated that in the case of the black cold-twisted deformed 410C bars the corrosion had taken place predominantly on the deformation ridges and opposite air voids, but in the case of the second bar, where the mortar was inhibited by 70-ppm sodium chromate, the surface of the bar and the mortar opposite were covered with a gray film. In the case of the black structural hot-rolled 230S grade bar, which was

TABLE 1—*Accelerated corrosion of reinforced mortar test cylinders: order of appearance of red rust staining.*

Concrete and Reinforcement	Time of Exposure, days	Total Current, Ah/m^2
No inhibition black 410C	4	34.5
70-ppm Na_2CrO_4 black 410C	6	52
No inhibition black 230S	11	97.5
70-ppm Na_2CrO_4 black 230S	14	120
No inhibition galvanized 230S	no rusting 21	180
70-ppm Na_2CrO_4 galvanized 230S	no rusting 21	180

the third to fail, the rusting was again on the ridges of the deformation but was more uniformly distributed than in the two previous cases, although a few isolated areas on the reinforcing bar opposite air voids were severely attacked. The fourth cylinder was inhibited by sodium chromate and, apart from the red rust on the ridges of the deformed structural 230S bar, the grey film previously noted was again evident. The mortar cylinders reinforced with galvanized structural-grade bars exhibited no evidence of rust after 21 days, but on opening the cylinder having no sodium chromate inhibition a thick layer of white corrosion products was evident on the surface of the galvanized bar while in some areas red oxide was noticeable. Finally, the mortar cylinder inhibited with 70-ppm sodium chromate was opened and similar white corrosion products were observed, but these were nonuniform and were present only where the grey film had been broken.

Apart from the cement mortar cylinders subjected to impressed anodic currents, three other concrete test blocks were cast, all having 16-mm deformed structural-grade bars as reinforcement. The first block contained 70-ppm sodium chromate, the second 70-ppm chromic oxide, while the third contained neither inhibitor. The intention of the experiment was to evaluate the physical difference between the two inhibitors on the control of hydrogen evolution at the reinforcement-concrete interface. After curing, the blocks were opened and it was observed that the concrete adjacent to the reinforcing bar, in the case where chromic oxide had been added, was quite porous although not so porous as the concrete which had not been inhibited. On the other hand, the concrete inhibited by sodium chromate exhibited no evidence of porosity and the interface was particularly smooth. All these bars had been hot dip galvanized.

Passivation of Zinc in the Presence of Inhibitor Anions

Potential as a function of time for a zinc electrode immersed in a saturated calcium hydroxide solution having inhibitor concentrations of 70 or 300 ppm of either sodium chromate or chromic oxide is shown in Fig. 1. All four of these curves exhibit an increase in potential with time. Furthermore, as the amount of sodium chromate or chromic oxide increased, the potential of zinc also increased and the potentials of zinc with the two additions of sodium chromate were higher than those for equivalent quantities of chromic oxide. It is important also to note that when sodium chromate was added a stable potential was reached, whereas with chromic oxide the potential continued to increase to prolonged times.

The plots of potential with respect to current shown in Fig. 2 are for zinc electrodes passivated in saturated calcium chloride with 70-ppm additions of either sodium chromate or chromic oxide. In both cases the zinc potential decreased sharply when the current was first applied, but

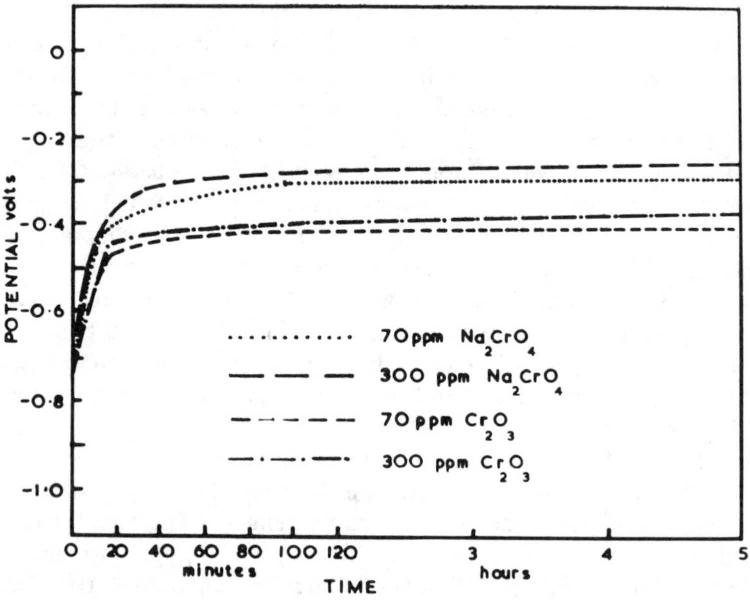

FIG. 1—*Potential of zinc versus time in saturated calcium hydroxide solutions inhibited with sodium chromate and chromic oxide.*

there was a greater loss of protection of the zinc in the solution inhibited with chromic oxide although the passivity became stable on further increases in current. On the other hand, the potential of zinc continued to decrease when inhibited with sodium chromate, eventually stabilizing but over a smaller current range.

Passivation of Zinc in the Presence of Both Inhibitor Anions and Aggressive Anions

The previous experiments were repeated using sodium chloride additions giving chloride ion concentrations of 10^{-3} M, 10^{-2} M, 10^{-1} M, and $10^{-0.5}$ M. In Figs. 3-6 a clean zinc electrode was used while in Figs. 7 and 8 an already passivated electrode was used. In the case of the previously passivated electrodes there was a decrease in potential, with the higher chloride ion concentrations of 10^{-1} M and $10^{-0.5}$ M in solutions inhibited with either sodium chromate or chromic oxide. On the other hand, in the case of the clean zinc electrodes, this decrease was observed only in calcium hydroxide solutions inhibited with chromic oxide. Irrespective of whether the zinc electrode was rendered active or passive at the higher chloride ion concentrations, the initial rate of change of the zinc potential decreased with increasing chloride ion concentration and it is noticeable that the presence

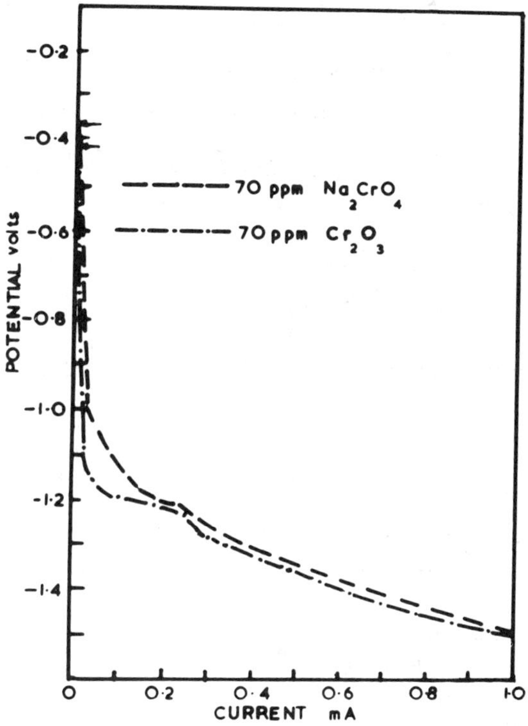

FIG. 2—*Potential of passivated zinc versus current in saturated calcium hydroxide solutions inhibited with sodium chromate and chromic oxide.*

of small amounts of chloride ions changed the shape of the plots quite considerably.

Figures 9 and 10 show a series of plots of potential measured with respect to current for zinc electrodes passivated in a saturated calcium hydroxide solution having a pH of 12.5 with 70-ppm sodium chromate or chromic oxide as an inhibitor and with a range of chloride ion concentrations.

Discussion

Corrosion Behavior of Reinforced Concrete Cylinders

The corrosion behavior of this series of reinforced cement mortar cylinders subjected to high anodic currents was intended to be assessed from the qualitative aspect of comparative exposure life, because the current density was far removed from the conditions of natural exposure and, moreover, the bar was forced to corrode. Nevertheless, a comparative

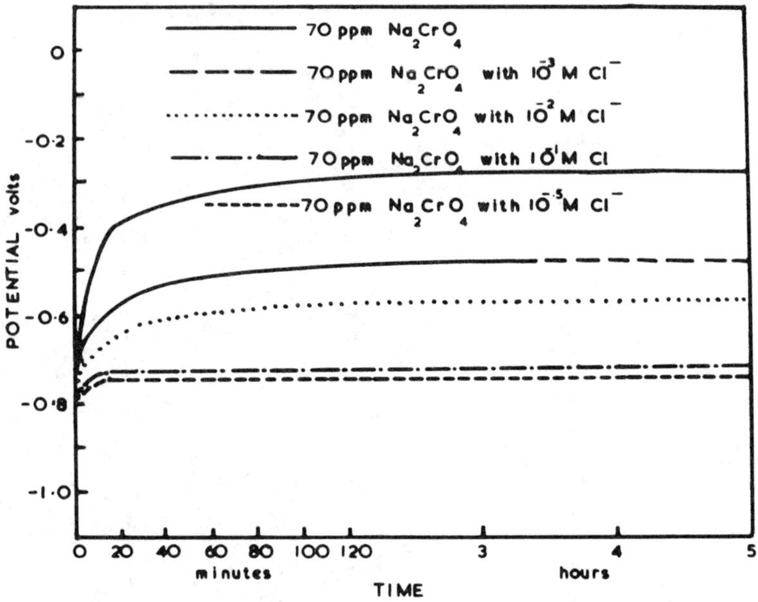

FIG. 3—*Potential of zinc versus time in saturated calcium hydroxide solutions inhibited with 70-ppm sodium chromate for different sodium chloride additions.*

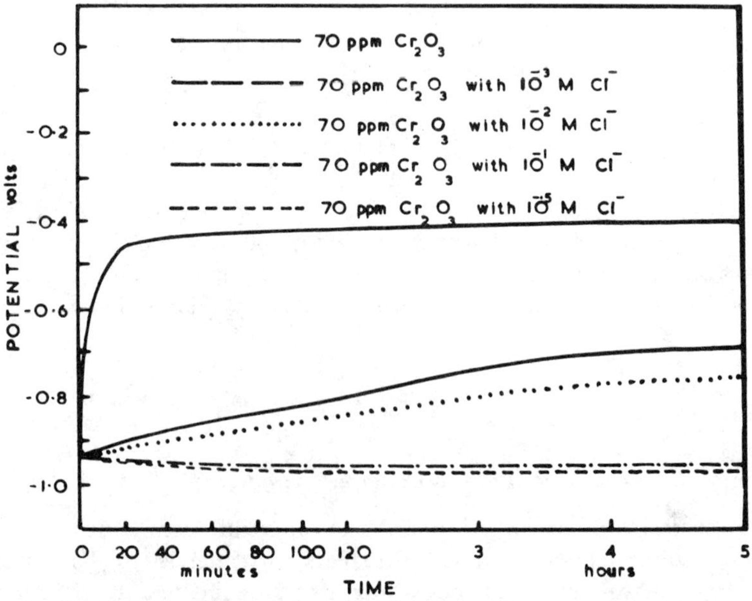

FIG. 4—*Potential of zinc versus time in saturated calcium hydroxide solutions inhibited with 70-ppm chromic oxide for different sodium chloride additions.*

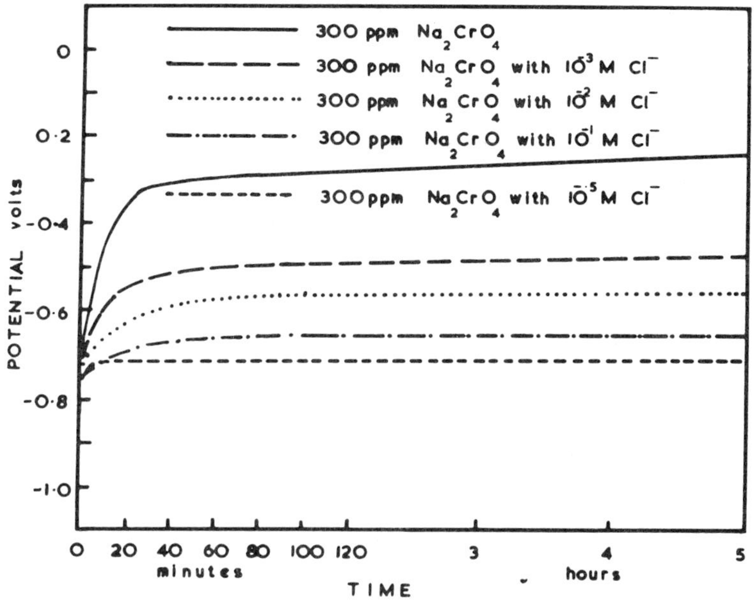

FIG. 5—*Potential of zinc versus time in saturated calcium hydroxide solutions inhibited with 300-ppm sodium chromate for different sodium chloride additions.*

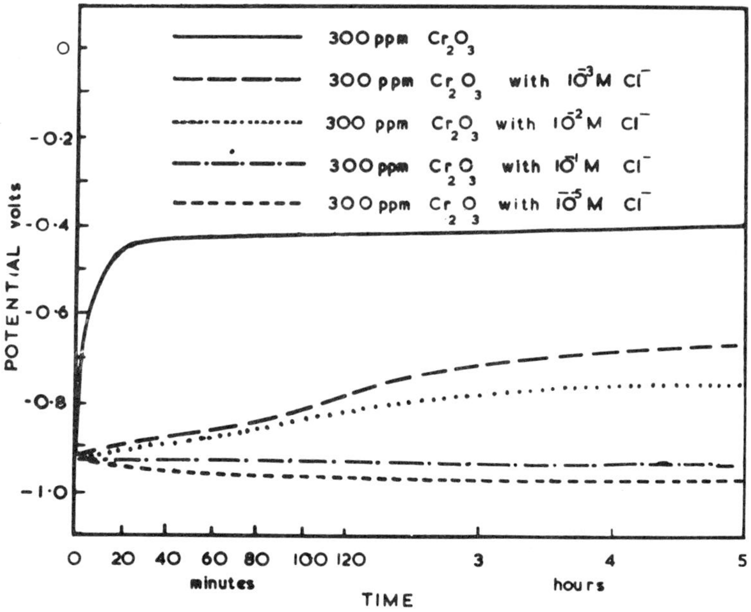

FIG. 6—*Potential of zinc versus time in saturated calcium hydroxide solution inhibited with 300-ppm chromic oxide for different sodium chloride additions.*

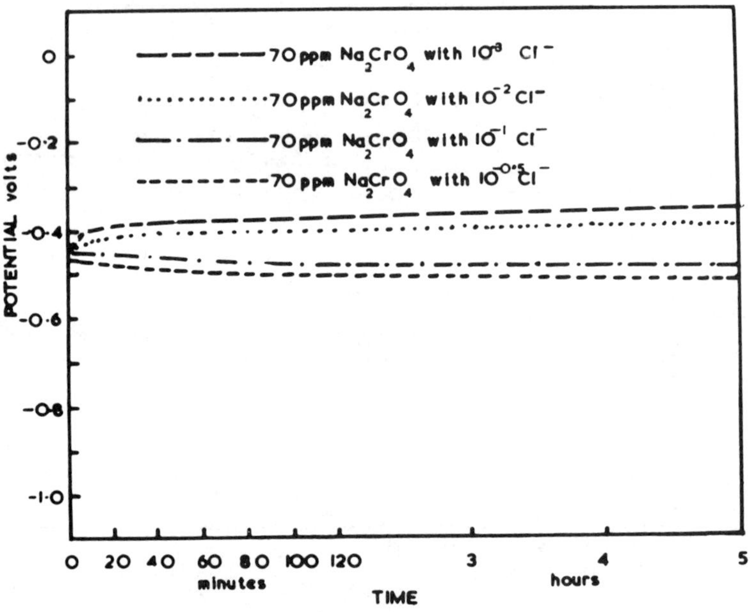

FIG. 7—*Potential of passivated zinc versus time in saturated calcium hydroxide solutions inhibited with 70-ppm sodium chromate for different additions of sodium chloride.*

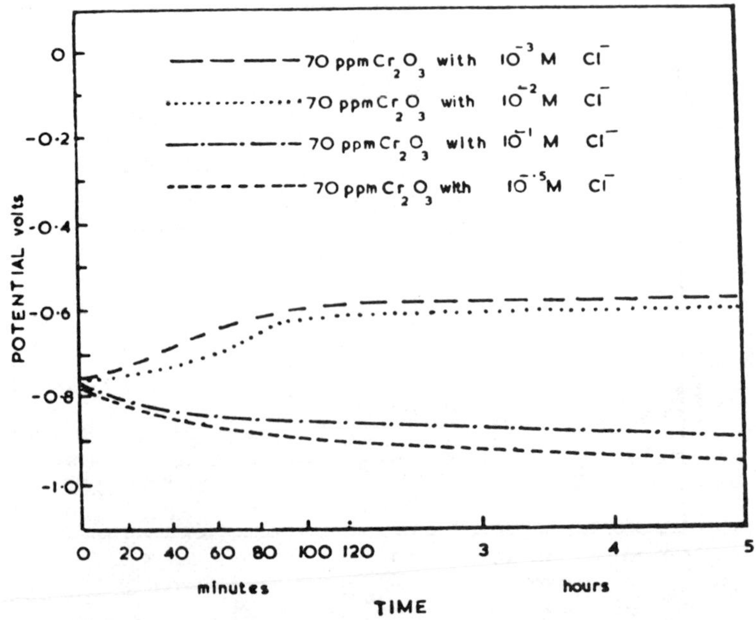

FIG. 8—*Potential of passivated zinc versus time in saturated calcium hydroxide solutions inhibited with 70-ppm chromic oxide for different additions of sodium chloride.*

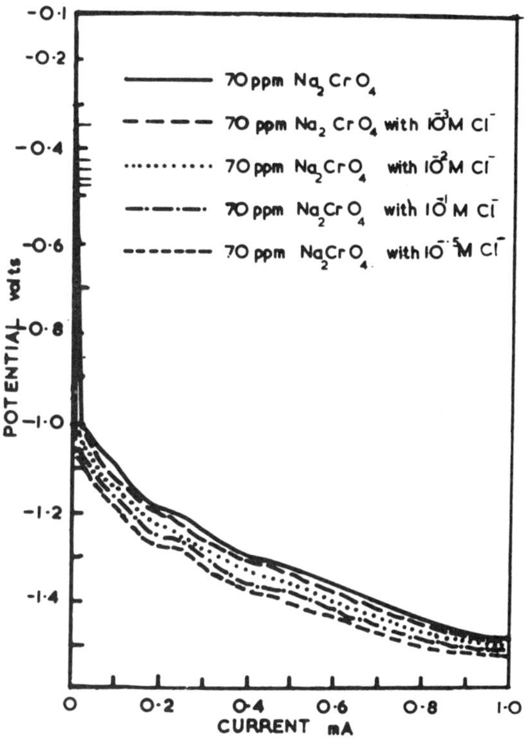

FIG. 9—*Potential of passivated zinc versus current in a saturated calcium hydroxide solution inhibited with 70-ppm sodium chromate for different sodium chloride additions.*

analysis showed a trend in the corrosion characteristics of reinforced mortar or concrete having different degrees of protection whether by way of a passivated film, galvanized surface, or by mill scale. This trend showed that as the protection of the bar increased, so did the uniformity of corrosion of the bar and the time taken for the bar to corrode.

The first mortar cylinder to exhibit rust stains on the concrete surface was reinforced with a cold-twisted deformed black Grade 410C bar which, on opening, showed excessive corrosion in a few isolated areas. Evidently, during the casting of the mortar, discontinuities of the bar-mortar interface had set up differential aeration cells. Other discontinuities were formed by mill scale, broken during the twisting process, not having flaked off completely. The mortar inhibited with sodium chromate showed more uniform corrosion on the bar surface and it took longer for the rust stains to reach the bar surface. Infrared spectroscopic analysis showed that the chromates in the mortar had been reduced to form a chromic oxide film. Nevertheless, some corrosion sites were present on the reinforcing bar where the passive film had been broken or not established. The mortar

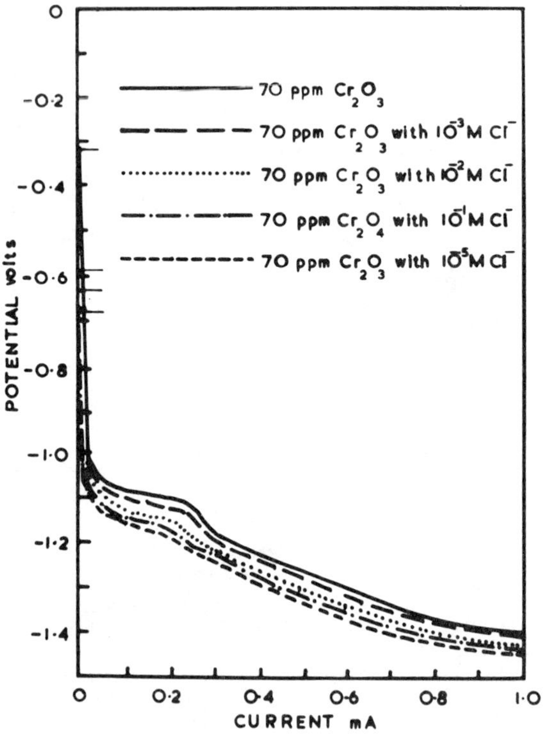

FIG. 10—*Potential of passivated zinc versus current in a saturated calcium hydroxide solution inhibited with 70-ppm chromic oxide for different sodium chloride additions.*

reinforced with the deformed structural 230S bar showed more uniform corrosion because it had the added protection of mill scale. The high impressed current density could not destroy the film as easily as in the case of the hydrated ferric oxide film. The trend continued in this way, with the deformed hot-rolled structural-grade 230S bar in the inhibited mortar giving the longest exposure of the black bars. The galvanized bars showed even more uniform corrosion. While no surface rust staining had occurred over the 21-day exposure period, white corrosion products had formed uniformly over the bar itself. In the case of the same bar in inhibited mortar, the white corrosion products had formed only where the passive film was broken. The white layer is attributed to the reaction of the zinc with calcium hydroxide which forms calcium hydroxo-zincate. The further gray layer which formed over the white corrosion products with chromate-inhibited mortar is thought to be a combination of zinc chromate, chromic oxide, and calcium hydroxo-zincate.

The preceding results are explained on the basis that when a reinforcing bar had little protection, differential pH cells and differential aeration cells were set up by discontinuities in the mortar opposite the reinforcing bar. The areas on which the cells formed were predominantly oxide-free regions at the bottom of accelerated cells. During an impressed current test, the passage of electrons was concentrated at these points, causing localized corrosion. As the protection of the reinforcing bar increased, however, these points became less apparent so that corrosion occurred more uniformly and at a considerably reduced rate.

Passivation and Corrosion Characteristics of Zinc

These experiments on the passivity of zinc were performed using both sodium chromate and chromic oxide in a calcium hydroxide solution with the intention of comparing the efficacy of the different inhibitor solutions. Both of these inhibitors have been recommended as additions to the concrete reinforced with galvanized steel to passivate the zinc and prevent the evolution of hydrogen at the zinc surface since chromates are reduced in preference to water. Aggressive halide ions in the form of chloride ions were added in different concentrations to study the effects of chloride ions on both passivated and clean zinc electrodes in the presence of inhibitor anions.

Reduction of chromates produces chromic oxide, and, when the zinc is oxidized, zinc chromate also forms. A small amount of chromate ion was determined in the calcium hydroxide solution inhibited with both 70- and 300-ppm chromic oxide. Thus, it would be expected that the passive layer formed on both electrodes would be composed of chromic oxide and zinc chromate. However, because of the smaller concentration of chromate present in the solution inhibited with chromic oxide, the passive film there would be weaker and the zinc would become more active. Both Duval et al [*2*] and Rehm et al [*3*] have shown that zinc is passivated in a calcium hydroxide solution by the formation of calcium hydroxo-zincate on the surface. This could also be true where the solution is inhibited with chromic oxide, in which case chromic oxide particles could be incorporated into the film to increase the corrosion resistance of the zinc. Both Figs. 1 and 2 support this analysis. Figure 1 shows the initial formation of passive films which form at the same rate in both solutions and could indicate the formation of a film composed of zinc chromate and chromic oxide. On this basis the zinc passivated in sodium chromate became more stable, since a greater concentration of chromate was present and a stronger passive film was formed. On the other hand, the zinc passivated in the solution containing chromic oxide was not as stable and the calcium hydroxo-zincate layer continued to grow. The respective breakdowns in passivity as demonstrated in Fig. 2 show that the passive film formed in the solution containing sodium chromate is the stronger of the two, while the stable regions in both plots indicate that the same type of film is being disrupted, presumably the zinc chromate-chromic oxide film.

Passivation and Corrosion Characteristics of Zinc in the Presence of Both Inhibitor and Aggressor Anions

In these experiments, the effect of increasing concentrations of aggressor anions in the form of chloride ions was studied using saturated calcium hydroxide solutions inhibited by either sodium chromate or chromic oxide. Figure 3 shows that a passive film continued to form in the presence of chloride ions but that the degree of passivity decreased with increasing chloride ion concentrations. With 70-ppm sodium chromate and a chloride ion concentration of $10^{-0.5}$ M, the zinc was passivated but only to a small extent. At the chloride ion concentrations of 10^{-3} M and 10^{-2} M it took 170 min for the zinc to stabilize, while for chloride ion concentrations of 10^{-1} M and $10^{-0.5}$ M the zinc stabilized after 15 min. This can be interpreted in terms of a competition between inhibitor anions and aggressive anions for the zinc surface. At low chloride ion concentration, the inhibitor anions were able to form a passive film, but chloride ions hindered this formation and further passivation of the zinc was restricted. As the chloride ion concentration increased, the passivity decreased, while the time required for a stable passive layer to form also decreased since the increased chloride ion concentration limited the further formation of a passive film. Gouda et al [*12*] have proposed that a critical concentration of inhibitor anions exists that can tolerate a range of concentrations of the aggressive anions. In the situation shown in Fig. 3, 70 ppm of sodium chromate provided a concentration of chromate ions greater than the critical concentration for the range of concentrations of chloride ions.

In the case of chromic oxide inhibitor, Fig. 4 shows that at low chloride ion concentrations the potential of the zinc electrode increased at a low rate with the zinc becoming passive, while at high chloride ion concentrations the zinc became more active with increases in times. This latter behavior indicates that the zinc chromate-chromic oxide film does not form at all and that concentration of chromate ions in the solution containing 70 ppm of chromic oxide is lower than the critical concentration for all concentrations of chloride ions. The increase in passivity at low chloride ion concentrations could be attributed only to the formation of a calcium hydroxo-zincate film. The increase in activity at chloride ion concentrations of 10^{-1} M and $10^{-0.5}$ M indicates that there exists a critical concentration of chloride ions above which zinc will not become passive in a calcium hydroxide solution of a certain pH. Alternately, there exists a critical pH that can tolerate a range of concentrations of chloride ions.

The results shown in Fig. 5 for the case of 300-ppm sodium chromate indicate that with this higher concentration of chromate a slightly higher degree of passivation was achieved because the effects of the chloride ions were lowered. However, the passivity was not much greater than that shown in Fig. 3 owing to the critical concentration having been already exceeded for that range of concentrations of chloride ions when the sodium chromate content was 70 ppm.

Figure 6 for chromic oxide concentrations of 300 ppm is very similar to Fig. 4 where the chromic oxide concentration was 70 ppm, because the chromate concentration, although 300 ppm, was still lower than the critical concentration and the pH of the solution containing 300 ppm of chromic oxide was the same as the pH of the solution containing 70 ppm chromic oxide.

The effect of a zinc electrode previously passivated in 70-ppm sodium chromate is shown in Fig. 7, which indicates that at chloride ion concentrations of 10^{-3} M and 10^{-2} M the corrosion resistance already established increased, while for chloride ion concentrations of 10^{-1} M and $10^{-0.5}$ M some passivity was destroyed. The degree of passivity obtained was, moreover, greater than for the corresponding clean zinc electrodes in the same chloride ion concentrations.

In Fig. 8 the saturated calcium hydroxide solution was inhibited with 70-ppm chromic oxide while the zinc electrode was previously passivated. Increases in potential are shown for chloride ion concentrations of 10^{-3} M and 10^{-2} M while decreases in passivity of zinc are shown in chloride ion concentrations of 10^{-1} M and $10^{-0.5}$ M. Higher degrees of passivity were obtained for chloride ion concentrations of 10^{-3} M and 10^{-2} M than was the case for the clean zinc electrode, while in the case of chloride ion concentrations of 10^{-1} M and $10^{-0.5}$ M the loss of passivity caused these zinc electrodes to become active. Once again, this indicates that either a critical pH exists which can tolerate a certain chloride ion concentration, or a critical chloride ion concentration exists which can destroy or prevent a passive film from forming.

The plots in Figs. 9 and 10 show that as the chloride ion concentration was increased, so was the ease with which the passive film could be destroyed. The region of the potential-current relationship where zinc established a stable state became less apparent with increasing chloride ion concentrations. This indicates that the thin zinc chromate-chromic oxide film formed during passivation was quite readily destroyed by increasing chloride ion concentrations.

Conclusions

Zinc coatings on galvanized reinforcement were passivated in calcium hydroxide solution by the formation of a calcium hydroxo-zincate film. Zinc was also passivated in the same solution containing sodium chromate or chromic oxide, respectively. In the case of sodium chromate, the chromate ions were reduced instead of water so that hydrogen evolution was prevented. Where chromic oxide was present rather than sodium chromate, a smaller concentration of chromate ions was present and some reduction of water occurred. At the same time, the zinc was passivated to a lesser extent by the formation of the zinc chromate-chromic oxide film although the calcium hydroxo-zincate which was also formed assisted in passivating the zinc surface.

Chloride ions in the presence of chromate ions compete for the zinc surface. It was found that a calcium hydroxide solution containing 70-ppm of sodium chromate had a chromate ion concentration greater than the critical concentration for a chloride ion concentration of $10^{-0.5}$ M. On the other hand, both 70- and 300-ppm additions of chromic oxide to the calcium hydroxide solution were less than the critical concentration for a chloride ion concentration of 10^{-3} M, and any passivity was due to the formation of a calcium hydroxo-zincate film. In both cases of chromic oxide additions of 70 and 300 ppm, zinc passivated at chloride ion concentrations of less than 10^{-1} M but became active at greater concentrations. It seems reasonable therefore to assume that a critical pH exists which can tolerate a certain range of chloride ion concentrations.

The electrochemistry experiments have been correlated with experimental casts of reinforced concrete using galvanized reinforcement. The results show that 70 ppm of sodium chromate effectively inhibits the evolution of hydrogen gas, whereas 70 ppm of chromic oxide is not as effective.

In assessing the wider significance of these results, however, it should be recognized that the highest chloride concentration level of $10^{-0.5}$ M corresponds to only 1.1 percent by weight of chloride ion in solution, whereas in concrete subject to deicing salts the chloride content could be higher by a factor of three or more and in these cases the levels of inhibition would need to be correspondingly higher.

Acknowledgment

The authors wish to acknowledge the award of an ILZRO Fellowship to H. Herzog in support of the present study.

References

[*1*] Corderoy, D. J. H., *Journal of the Institute of Engineers, Australia,* Vol. 49, No. 15, 1977, pp. 19–20.

[*2*] Duval, R. and Arliquic, G., *Memoires Scientifiques de la Revue de Metallurgie,* Vol. 71, No. 11, 1974, p. 719.

[*3*] Rehm, G. and Lammke, A., *Betonstein Feitung,* Vol. 360, 1970, p. 36.

[*4*] Bird, C. E., *Nature,* Vol. 194, No. 4830, 1962, p. 798.

[*5*] Bird, C. E., *Corrosion Prevention and Control,* Vol. 11, No. 7, 1964, p. 17.

[*6*] Zinc Institute Inc., *Galvanized Reinforcement for Concrete,* International Zinc Lead Research Organization, New York, 1974.

[*7*] Cook, A. R. in *Proceedings,* International Association for Shell Structures Conference, June 1976, pp. 16-18.
[*8*] Mahaffey, P. J., Cement and Concrete Association of Australia Internal Report TR3/C/7407.
[*9*] Bracket, M. and Raharinaivo, A., International Lead Zinc Research Organization Project ZE-154 Report, June 1971.
[*10*] Kaesche, H. and Werkstoffe, V., *Korrosion,* Vol. 20, 1969, p. 119.
[*11*] Cornet, I. and Bresler, B., *Materials Protection,* Vol. 5, No. 4, 1966, p. 69.
[*12*] Gouda, V. K. and Sayed, S. M., *Corrosion Science,* Vol. 13, 1973, p. 841.

I. Cornet[1] *and B. Bresler*[1]

Critique of Testing Procedures Related to Measuring the Performance of Galvanized Steel Reinforcement in Concrete

---

REFERENCE: Cornet, I. and Bresler, B. **"Critique of Testing Procedures Related to Measuring the Performance of Galvanized Steel Reinforcement in Concrete,"** *Corrosion of Reinforcing Steel in Concrete, ASTM STP 713,* D. E. Tonini and J. M. Gaidis, Eds., American Society for Testing and Materials, 1980, pp. 160–195.

ABSTRACT: Laboratory investigations of the corrosion performance of galvanized steel and of black steel in concrete show inconsistencies, poor correlation between field performance and accelerated laboratory tests, and even a lack of theoretical justification and understanding of modeling principles in testing.

This critique examines factors which influence the relationships between test results and prototype performance: loading, environment, geometry, and materials. Laboratory investigations of corrosion of black and of galvanized steel in concrete are reviewed. Bond characteristics, ductility, strength, and fatigue strength of galvanized reinforcement are presented. Field performance is discussed.

Emphasis is placed on planning of field and laboratory tests. Four primary requirements are proposed: structural integrity, functional integrity, durability, and aesthetic appearance. There are no standardized or widely accepted test procedures to evaluate corrosion performance of reinforcing or prestressing steel in concrete exposed to aggressive environments.

KEY WORDS: testing procedures, corrosion performance, galvanized steel reinforcement in concrete, accelerated tests

Galvanizing has widespread use as a means of providing corrosion protection for steel reinforcement in concrete exposed to aggressive environments. In the past 20 years, there have been many investigations of the corrosion performance of such reinforcement (ungalvanized and galvanized), including research on the corrosion mechanisms as well as corrosion measurement studies.

[1] Professors emeritus, Department of Mechanical Engineering and Department of Civil Engineering, respectively, University of California, Berkeley, Calif. 94720. Coauthor Bresler is also principal, Wiss, Janney, Elstner and Associates, Inc. Emeryville, Calif. 94608.

In devising testing procedures, two types of test may be considered:

1. Given a structure in a specified environment, how can a laboratory model test be devised which will be representative of the prototype structure?
2. How can results of standard tests on relatively simple specimens exposed to controlled environment be interpreted to provide indications of performance of a variety of structures in different environments?

The important factors which influence the relationships between test results and prototype performance fall into four primary categories: loading, environment, geometry, and materials. These factors are briefly discussed in the following.

Loading—In many cases, the level of stress and cyclic loading will affect the initiation, growth, and distribution of cracks in the concrete cover, thus influencing subsequent corrosion performance. Very few tests, to date, have incorporated stress or cyclic stress on the specimen concurrent with exposure to aggressive environment.

Environment—The normal aggressive environment may be accentuated, but there should be some basic understanding of corrosion mechanisms to guide the design of an accelerated test. Thus 3½ to 4 percent aqueous sodium chloride solution or seawater may be a reasonable exposure medium where a marine atmosphere is the environment to be modeled, but for most exposures saturated sodium chloride solution is an unreasonable environment as it is actually lower in oxygen solubility than the natural salt solutions encountered.

Tests combining aggressive environment with freeze-thaw cycling, heating and cooling cycling, and elevated temperature may be important and informative, but very few investigations have covered such temperature effects, presumably because of the expense.

Calcium hydroxide solutions have often been used to simulate the chemical factors in concrete, but physical factors such as convection and diffusion are markedly different in $Ca(OH)_2$ solution as compared with the solid concrete.

Ice removal procedures involve application of concentrated brine or solid salts such as sodium chloride or $Ca(Cl)_2$, but more or less dilution will occur before the chlorides penetrate into the concrete.

Impressed-current tests provide another means of accelerating an aggressive environment. In many structures, such as pilings, pipelines, and bridge decks, corrosion proceeds due to the presence of local anodes (where steel is corroding) and local cathodes (where oxygen is being reduced). These anodes and cathodes can be close together or some distance apart. The differences in potential involved may be as much as half a volt due to differences in both oxygen and chloride concentration. Impressed-current tests may, therefore, be reasonable types of acceleration, provided the impressed potentials and currents are controlled to avoid hydrogen liberation, oxygen liberation, passivation, etc.

Geometry—The size and spacing of steel reinforcement and the thickness of cover are important parameters, but cannot be scaled linearly with respect to the prototype. Investigators rarely attempt to relate the geometry to the degree of acceleration quantitatively, but often draw conclusions from tests which are not properly related to the prototype.

If a reinforced-concrete specimen shows distress in two years with 12.7-mm (½ in.) cover, how long would a structure with 50.8 mm (2 in.) of cover be expected to resist the same aggressive exposure? If the distress is a function of diffusion of oxygen and chloride ion, quantifying results may be feasible [*1,2*].[2] Diffusion models for corrosion in slabs, prisms, and cylinders lead to quantitative relationships which should be considered in setting up tests.

Materials—A common form of acceleration is to vary materials. Test specimens often use a water/cement (W/C) ratio of 0.6 or 0.7 where an actual structure would have concrete with W/C = 0.45 or 0.5. No reliable relationship between W/C ratio and corrosion protection has been established. Also, the initial steel surface condition can be important. Loose rust on smooth round reinforcement may be more susceptible to further corrosion and give poor bond, while a rusted pitted surface on plain bar may give good bond. A deformed reinforcing bar which has been sandblasted for epoxy coating may show improved fatigue performance compared with a bar that was not sandblasted, because sandblasting removes some of the sharp notches at the lugs. A galvanized steel reinforcing bar may show better corrosion performance in concrete with a hexavalent chromium content of 100 ppm based on cement content than in concrete with much less chromate—and normally one does not test or report on such trace chemicals as chromium in cement.

Sometimes in a prestressed steel slab structure there will be delayed fracture of the tendons where post-tensioning steel cuts through sulfide-containing aggregate, although tests showed the steel, as received, good, and the concrete passed normal strength requirements.

Design of appropriate testing procedures relating to corrosion performance of steel reinforcement in concrete should account for the four major factors just reviewed and draw on the following three types of available information:

1. Results of laboratory investigations dealing with corrosion performance.
2. Results of laboratory investigation dealing with bond and material characteristics related to performance of galvanized steel reinforcement in concrete.
3. Observations and measurements of field performance related to full-scale structures.

[2] The italic numbers in brackets refer to the list of references appended to this paper.

A critical review of selected papers dealing with these topics is presented in the next section.

Laboratory Investigations of Corrosion in Reinforced Concrete

General

Various methods of corrosion testing have been used. By 1964, there were at least 15 types of corrosion tests used on black steel in concrete specimens exposed to approximate service conditions with (more or less) acceleration of corrosion. There were exposures to (1) tidal water (alternate immersion) [*3,4*]; (2) normal outdoors [*4-7,20,51*]; (3) laboratory-high humidity [*5,7-9*]; (4) laboratory-low humidity [*5,10*]; (5) alternate immersion [*5,9,11*]; (6) alternate but partial immersion [*4,13*]; (7) variable salt, moisture, and temperature [*14*]; (8) salt spray cabinet [*15*]; (9) covered with wet towel [*10*]; (10) flow of water vapor [*16-19*]; (11) periodic spraying with salt water [*16-18*]; (12) immersed in water [*9*]; (13) partial immersion [*20*]; (14) dry cellar [*21*]; and (*15*) impressed voltages [*22*].

More recent variations of exposure systems include (16) partial immersion in saturated sodium chloride solutions [*23,24*]; and (17) intermittent or periodic application of deicing salts [*25-27*].

Most of the aforementioned tests used specimens which were relatively small, with a steel wire or reinforcing bar, black or galvanized, embedded completely or partially in a concrete block. A few tests have included beams, slabs, wall panels, and rectangular prisms of greater size. There have been a few tests in which specimens have been subjected to approximately working loads while exposed to the corrosive environment, and some tests with cyclic loading combined with exposure.

Generally, laboratory test specimens of concrete with galvanized reinforcement included similar specimens with black or with epoxy-coated specimens. Results have then been expressed in relative terms. A few such tests have been selected for detailed discussion.

In considering use of galvanized reinforcement in concrete, one often encounters the remark that zinc is an amphoteric metal which reacts with an alkaline environment. The data of Roetheli et al [*28*] are widely cited in corrosion textbooks. These data were obtained for small flat pure zinc specimens exposed to aerated aqueous solutions of hydrochloric acid and of sodium hydroxide at 30°C (86°F) for test periods varying from 5 to 30 days so that the same weight loss was obtained. The data have been interpreted as indicating a very high corrosion rate for zinc in concrete. Plotted in Fig. 1, the same data clearly indicate a minimum in corrosion rate of about 0.005 cm/year (2 mils/year) at a pH of 12.4, which is characteristic of cured concrete. However, these data are for aerated sodium hydroxide

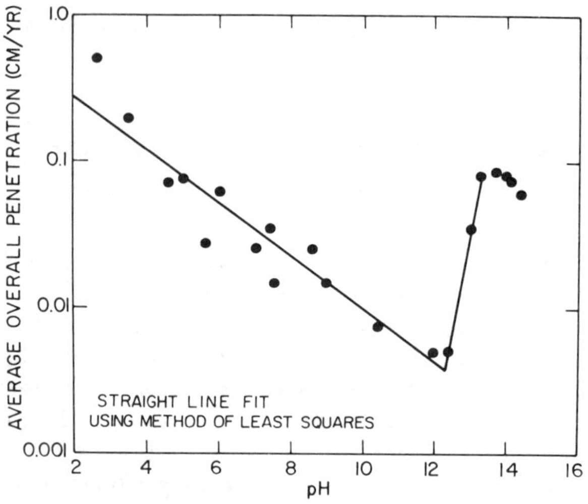

FIG. 1—*Corrosion of zinc in agitated aerated solutions of hydrochloric acid and sodium hydroxide* [28].

solutions with agitation, and represent penetration from both sides of a sheet specimen.

Saturated calcium hydroxide solution is often used as a test solution to simulate the chemical environment of concrete. It has been established that zinc corrodes in this environment to form calcium hydroxyzincate, an adherent film which acts as a diffusion barrier to stifle further corrosion [*29,30*]. Furthermore, hexavalent chromium compounds (CrO_3) in concentrations of about 70 ppm based on mixing water (about 50 ppm based on weight of cement) passivate the zinc and inhibit the reaction of zinc with hydroxyl ions to form hydrogen gas [*31*]. The data of Ref *28* are, therefore, not particularly relevant to galvanized reinforcement in concrete.

Bresler, Cornet, and Ishikawa

Among the early reports on corrosion performance of galvanized reinforcement in concrete are papers by Bresler and Cornet [*11*], Cornet and Bresler [*32,33*], Cornet et al [*34,50*], Cornet [*2*], and Ishikawa et al [*35*]. Their studies were performed principally with rectangular concrete prism specimens, 10 by 10 by 30 cm (4 by 4 by 12 in.), reinforced with 1.9-cm-diameter (0.76 in.) bar 122 cm (48.8 in.) long. They used plain and deformed bars, black and galvanized. A center notch 1.3 cm (0.52 in.) deep and 0.16 cm (0.064 in.) wide was cut into the concrete prism after 21 days' curing and before exposure. They exposed specimens under a working load of 142 MPa (20 000 psi) nominal stress on the reinforcement (1) in laboratory air; (2) in alternate immersion—4 percent sodium chloride solution for three days alternating with four days for drying; and (3) in a 4 percent sodium chloride solution with impressed anodic current [215 mA/m^2 (20 mA/ft^2)] of reinforcement surface, at ambient temperatures of 24 to 29°C (75 to 84°F). They observed the initiation and growth of cracks, and measured crack lengths and widths and slip across the notch.

The Bresler and Cornet data are presented in Fig. 2 [*33*]. The bond between the concrete and the reinforcement decreased more rapidly with black steel than with galvanized steel. The data indicate that under corrosive conditions concrete cracked earlier, and the cracks grew longer

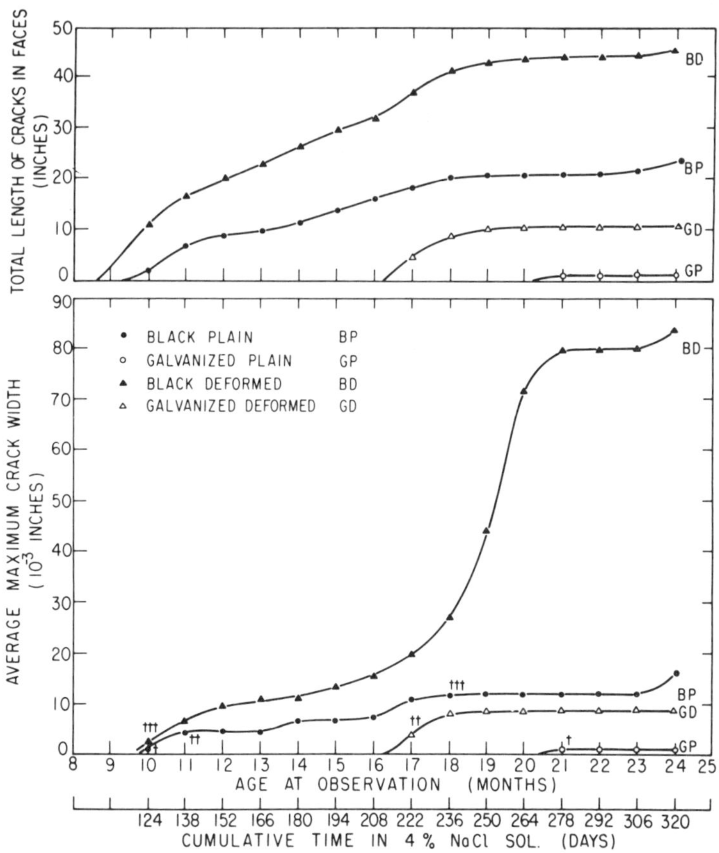

Maximum width and total length of cracks are averages for three specimens. Age at which different specimens cracked and total number of specimens cracked are shown by †: one specimen †, two ††, three ††† in lower figure only.

FIG. 2—*Corrosion performance under alternating immersion in salt solution (1 in. = 2.54 cm).*

when the reinforcement was black as compared with galvanized steel. Figure 3 shows specimens used by these authors.

It is worth noting that with all of the care in preparing and exposing specimens in this study, there may still be about a 2-to-1 variation in performance. For example, of black bar specimens subjected to alternate immersion, four cracked after 9 months, a fifth cracked after 10 months, and a sixth cracked after 18 months' exposure, as shown in Fig. 2 [*33*].

Griffin

Griffin [*16–18*] used concrete walls 87.6 cm (34.5 in.) high, 91.4 cm (36 in.) wide, and 8.9 cm (3.5 in.) thick, reinforced with No. 5 steel bars. He used a 2.54-cm (1.0 in.) cover of concrete, with water/cement ratios of 0.70 (nonair-entrained) and 0.64 (air-entrained), with cement content of 4.8 sacks of cement per cubic yard (3.6 sacks/m^3) of concrete. Reinforcement variables included uncoated or galvanized steel grid walls without reinforcement, and either welded or insulated and tied grid connections. Walls were sprayed once daily with seawater for about three years in outdoor exposures at Port Hueneme, Calif. Data are presented in Table 1. Three

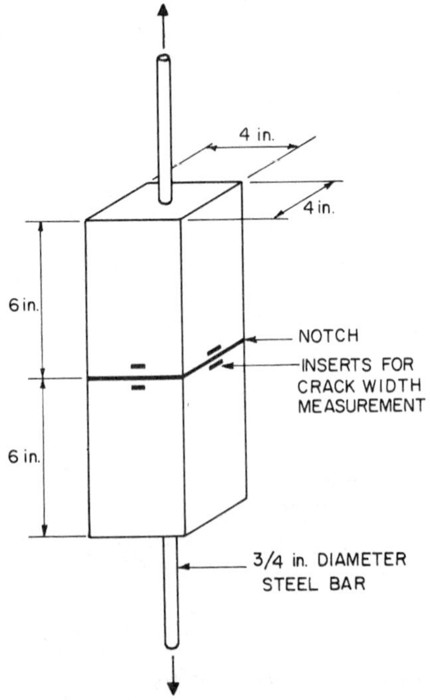

FIG. 3—*Corrosion specimen details (1 in. = 2.54 cm).*

TABLE 1—*Results of wall tests (after Griffin)* [16-18].

Wall No.	Type of Steel	Air Entrainment	Time to First Cracking, days
2[a]	uncoated	no	485 (R306 Supplement)
101	galvanized	no	570
109	galvanized	no	585
111	galvanized	no	634
104	uncoated	no	722
103	galvanized	no	809
106	uncoated	no	991
105	uncoated	yes	no cracks reported
102	galvanized	yes	no cracks reported
110	galvanized	yes	no cracks reported

[a] Wall No. 2, similar to Walls 104 and 106, is reported by Griffin in Ref *17*.

of the nine walls of the 1969 report used air-entrained concrete. No cracks were reported for any of these for the three years of exposure, and they included one wall with uncoated and two with galvanized reinforcement. The walls without air entrainment all exhibited cracking during the exposure period. These included four with galvanized and two with uncoated steel. Data on Wall No. 2 have been added to the 1969 data, because Wall No. 2 is comparable to Walls 104 and 106, having the same thickness of cover, W/C ratio, cement factor, etc.

Results obtained on a small number of concrete specimens are always subject to question, particularly where time-dependent processes are involved. Walls 104 and 106 exhibited cracks at 722 and 991 days, while in Griffin's earlier study [*16*] a similar panel, Wall No. 2, exhibited stains and cracks at 485 days. This 2-to-1 ratio is in accord with the findings mentioned in the foregoing [*33*]. Drawing conclusions on the basis of these limited data is questionable.

A detailed review of Griffin's results raises other questions. For example, five walls had early cracks in the east end face and only one in the west end face. This suggests the possibility that some parameter other than the coating on the steel might be involved.

Reinforcement of Walls 101, 103, and 104 were examined in some detail. There was evidence of white zinc corrosion products, with some scattered red rust spots, indicating that zinc was corroding preferentially and protecting the steel. The uncoated steel reinforcement showed more extensive and severe rusting, with black to red oxide deposits over the entire length.

Lorman

Lorman [*36*], after a critical appraisal of the technical literature dealing with thin wall reinforced concrete exposed in marine conditions for the period covering the past 75 years, concluded that the use of galvanized reinforcement for thin wall reinforced concrete floating piers would be beneficial.

Clear and Hay

Clear and Hay [*25*] and Clear [*26*] used concrete slabs 1.2 m by 1.5 by 15.2 cm (4 ft by 5 ft by 6 in.) thick with a grid of No. 4 steel reinforcing bars. They used concrete with W/C ratios of 0.40, 0.50, and 0.60; a cement factor of 390 kg/m^3 [658 lb/yd^3 (7.3 sacks)]; and cover thickness of 2.54, 5.08, or 7.62 cm (1, 2, or 3 in.). Slabs were exposed to outdoor conditions and one surface of each slab was subjected to a ponding depth of 0.15 cm ($1/16$ in.) of 3 percent sodium chloride solution each afternoon for 330 consecutive days [*25*] and 830 days [*26*].

Half-cell potentials of black reinforcing steel were monitored at weekly intervals at six positions on the slabs. Potentials consistently greater than −0.35 V versus copper-copper sulfate electrode (CSE) were taken to indicate a high probability of corrosion. Potentials consistently less numerically than −0.20 V versus CSE indicated a high probability of no corrosion. Potentials between −0.20 and −0.35 V versus CSE were considered to be uncertain.

Chloride content was measured by dry coring and analysis. The threshold concentration required to initiate corrosion of steel in portland cement concrete was assumed to be 330-ppm chloride ion [0.77 kg chloride ion per cubic metre of concrete (1.3 lb chloride ion per cubic yard of concrete)].

There was visual observation for signs of concrete distress. In their 1973 report [*25*], Vol. 1, Clear and Hay describe 124 slabs, including four with galvanized steel reinforcement; in Vol. 3 [*26*], Clear describes 120 slabs, omitting discussion of the performance of slabs with galvanized steel reinforcement. In Vol. 1 Clear and Hay report no visible concrete deterioration occurring in slabs with galvanized reinforcement. It is not reported whether these tests were continued; it was stated that results for slabs with galvanized steel reinforcement in concrete were inconclusive, apparently due to lack of a known threshold value of chloride ion for initiation of corrosion and due to uncertainty of a potential value for corrosion of galvanized steel in concrete.

Clear and Hay were principally concerned with the corrosion of steel in bridge decks. Their work showed very clearly the importance of concrete quality and cover in combating corrosion. They recommended a clear cover thickness of 5.08 cm (2 in.) with concrete having a W/C ratio of 0.40, and 7.62-cm (3 in.) cover with concrete having a W/C ratio of 0.50; they also recommended consolidation control of concrete with a slump less than 7.62 cm (3 in.). They did not make a conclusive evaluation of the effectiveness of galvanized reinforcing steel, because from their viewpoint it was susceptible to corrosion at the chloride concentrations obtained. They did not determine the half-cell potential or the chloride concentration at which galvanized reinforcement would corrode.

Their data have been extrapolated in a way which is appealing, but may not be reliable. It was proposed that 830 daily salt applications might be equivalent to over 40 years of service in an area which sees perhaps 20 storms a year in which salt is applied for deicing. However, the slabs were not subjected to cyclic loading corresponding to the effects of traffic on the bridge deck, and the effects of time intervals between saltings and seasonal changes in temperature, relative humidity, and rainfall remain uncertain.

The Federal Highway Administration (FHWA) has reported further on this staff research. Slabs were examined after 3.0 and 4.3 years of test, approximately 1000 and 1450 saltings, respectively. Concrete slabs with 0.5 W/C have far more deterioration than those with 0.4 W/C. Control slabs with the black steel reinforcement are cracked somewhat more visibly and frequently than the corresponding slabs with galvanized steel. Concrete slabs with W/C = 0.4 and galvanized steel reinforcement (only two slabs) are cracked, whereas the slabs with black steel are not. There were only four galvanized slabs in this study, a number insufficient to reach reliable conclusions.

Sopler

Sopler [*37*] and Bernhardt and Sopler [*38*] used 15 by 15 by 100-cm (6 by 6 by 40 in.) prism specimens, 202 in number, some of which were reinforced with four main steel bars and five stirrups. The main steel reinforcement consisted of 10-mm (4 in.) deformed bars with a yield strength of 40 kips/mm^2, while the stirrups were 6-mm (2.4 in.) plain round bars of mild steel. Three different concrete cover thicknesses (1.5, 3.0, and 4.5 cm) were investigated with three different types of concrete: W/C = 1.0, 0.67, and 0.50 with cement contents of 213, 305, and 394 kg/m^3 (359, 514, and 664 lb/yd^3), respectively.

Some of the reinforcement was galvanized; some was galvanized and chromated; and some was black.

Specimens were exposed to seawater 24 h after casting. Since this investigation was concerned with corrosion of concrete in a tidal zone, specimens were placed in tanks with a pump arrangement so that seawater level fluctuated over the middle 60 cm (24 in.) of prism heights. There were low-tide and high-tide periods of 4 h, each separated by pumping intervals of 2 h. Salt concentration of the seawater was 3.3 percent. Temperatures of exposure varied, with groups exposed to −5 to −8°C (23 to 18°F), +2 to −2°C (36 to 28°F), and +5 to +8°C (41 to 46°F). Time of exposure varied for groups from 30 to 88 months, with hot-dip galvanized specimens being removed at 40 and at 51 months. Electrical resistance and potential were measured as a function of time, and specimens were split and examined after exposure.

The effect of galvanizing was quite clear. Corroded areas were considerably smaller in the case of galvanized reinforcement compared with ordinary reinforcement. Pits were fewer in number and shallower with galvanized steel.

The combination of black with galvanized steel showed that there was protection of black steel stirrups where the main bars were galvanized. Where the stirrups were galvanized and the main bars were not, corrosion patterns resembled those of all-black steel systems. Judging by the degree of pitting and by the corroded areas, Bernhardt and Sopler concluded that hot-dip galvanizing will delay attack on steel in concrete in a tidal seawater exposure. In time, however, the concrete cracks to the same degree, whether the steel is bare or galvanized. Measurement of electrical potentials on the surface of the concrete indicated when the chloride ion reached the reinforcement. Electrical resistance measurements indicated the progress of the degree of hydration.

Relevant data of the Sopler report [*37*], reproduced here as Tables 2 and 3, indicate that galvanizing reinforcement retards rust spots and crack formation on concrete specimens.

The corrosion index, Table 2, is highly sensitive to quality of concrete (W/C) and thickness of cover. Galvanizing improves corrosion performance (lowers index) in all cases. It is particularly effective for concrete with a W/C ratio of 0.5 and a thickness of cover of 3.0 cm (1.2 in.). Cracks and rust spots are highly sensitive to thickness of cover and quality of concrete; Table 2 shows no cracks or rust spots on Specimens 35, 51, and 47, even though examination of embedded steel in Specimen 35 shows significant

TABLE 2—*Corrosion index of longitudinal bars* (L) *and stirrups* (S) [37].

Water/Cement Ratio		0.5				0.67			
	Concrete cover, cm	3.0		1.5		3.0		1.5	
	Location	L	S	L	S	L	S	L	S
I	All black steel (reference)	#35		#34		#32		#31	
		2.5	2.5	2.5	2.5	2.5	2.5	2.5	2.5
II	All galvanized steel without chromium	#51		#50		#49		#48	
		0	1.5	1.5	1.5	0.5	1.5	1.5	2.5
III	All galvanized steel with chromium	#47		#46		#45		#44	
		0	0.5	0.5	1.5	0.5	1.5	1.5	2.5

NOTE: The corrosion index tabulated here refers only to the condition of the steel: 0: no corrosion; 0.5–1.0: surface corrosion; 1.5–2.0: pitting; 2.5–3.0: serious pitting.

TABLE 3—*Rust spots* (RS) *and cracks* (CR) *on concrete specimens at various ages* [37].

Water/Cement Ratio		0.5				0.67			
Concrete cover, cm		3.0		1.5		3.0		1.5	
Observation		RS	CR	RS	CR	RS	CR	RS	CR
I All black steel (reference)		#35		#34		#32		#31	
On	Months								
May '70	19	...	...	1	1	...	...	11	...
Mar. '71	29	...	...	1	1	...	...	11	...
May '72	43	...	...	5	2	1	...	11	1
Feb. '73	52	...	...	6	2	1	...	11	2
II All galvanized steel without chromium		#51		#50		#49		#48	
	Months								
May '70	19	...	...	...	...	...	...	...	...
Mar. '71	29	...	...	...	...	...	...	...	...
May '72	43	...	...	...	...	...	...	...	...
Feb. '73	52	...	...	...	3	...	...	...	3
III All galvanized steel with chromium		#47		#46		#45		#44	
	Months								
May '70	19	...	...	...	...	...	...	...	...
Mar. '71	29	...	...	...	...	...	...	...	...
May '72	43	...	...	...	...	...	1	...	4
Feb. '73	52	...	...	1	2	...	2	4	4

presence of corrosion. Specimens with galvanized steel reinforcement show no rust spots until about 40 months' exposure; the number of cracks is about the same as for reference specimens.

The data, taken from Table 7.5.3 on pages 54 and 55 of the Sopler report, contradict Sopler's statements on pages 7, 48, and 49 of the same report that galvanizing induces strong crack formation in concrete. Thus, the Sopler data are consistent with the findings of Bresler and Cornet [*11,33*] that galvanizing the reinforcement delays cracking of concrete and reduces the number and size of cracks.

Hill et al

Hill et al [*24*] reported on laboratory corrosion tests of galvanized steel in concrete. They used prisms 11.4 by 6.3 by 38.1 cm (4.5 by 2.5 by 15 in.), each with a single steel reinforcing bar of 1.27 cm (½ in.) diameter partially embedded lengthwise in the prisms. Concrete specimens were either steam cured or moist cured, with W/C ratios of 0.72, 0.59, 0.47 and cement content of 2.96, 3.55, and 4.44 sacks/m^3 (5.0, 6.0, and 7.5 sacks/yd^3). They used 2.54 cm (1 in.) minimum thickness of cover. The 38.1-cm-long (15-in.) specimens were subjected to partial immersion [to a depth of 8.9 cm (3½ in.)] in saturated sodium chloride solutions for up to 1700 days.

Data are presented in Fig. 4. Black steel is designated by zinc of 0 oz/ft^2 in this figure. Examining the data for the 5-sack-mix moist-cured specimens, one finds that specimens with galvanized reinforcement take about 80 percent longer to crack than do specimens with black reinforcement. There is a 1.8-to-1, almost 2-to-1, improvement in performance displayed by galvanized reinforcement in concrete, based on values reported by Hill, Spellman, and Stratfull (H, S&S). Galvanized specimens show a 2.1-to-1 improvement in performance over black for steam-cured 5-sack concrete mix specimens.

The data for the 7½-sack mix are somewhat more complicated to analyze.

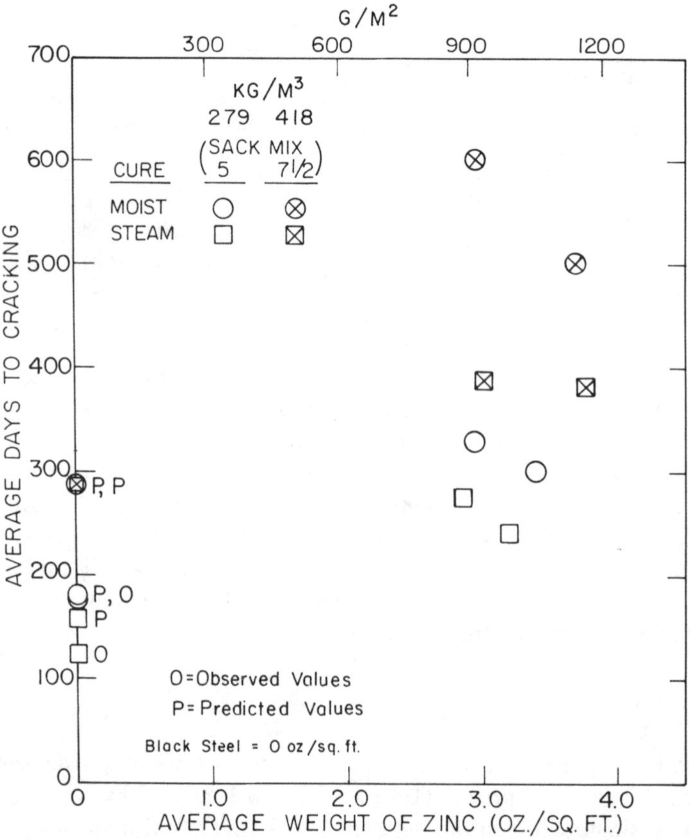

FIG. 4—*Time-to-cracking versus weight of zinc. Data predictions of Table 4 are shown above. For galvanized: each observed point represents 10 specimens. For black: each observed point represents 20 specimens. Data by Hill et al* [24] (*1 kg/m^3 = 1.685 yd^3; 1 oz/ft^2 = 0.3051 kg/m^2*).

Only seven out of 20 black steel specimens had cracked when testing was terminated, and no data are given for when those cracked. To permit detailed analysis, it is necessary to refer back to an earlier, 1968, report by Spellman and Stratfull [*23*].

Their data provide a prediction equation, $C = 1.12P+115$, where C is average days to concrete cracking and P is the average days to active potential. They used −0.35 V-SCE as the criterion for an active potential for steel in concrete in saturated sodium chloride solution. They state that "The effect of the orientation of the steel to the time to an active potential appears to be relatively minor," referring to whether the steel is cast vertically or horizontally in the concrete. In fact, they gave an equation, $D_v = 0.988D_h + 22$, where D_v is the average days to an active potential for a set of five specimens, steel cast vertically, and D_h is the average days to an active potential for a set of five specimens, steel cast horizontally.

Furthermore, Ref *23* used −0.35 V-SCE as an active half-cell potential for black steel in concrete, and that is what is used in calculating the predicted days to cracking tabulated in Table 4 and shown in Fig. 4 and 5. In their closure to Ref *24*, the authors state that the active potential of steel versus SCE is −0.27 V, but that is not what Ref *23* says. The time to an active potential of −0.27 V-SCE is shorter than the time to an active potential of −0.35 V-SCE except for the case where the potential-versus-time curve is almost vertical. It is estimated that the difference is no more than 15 days between the two potentials, which would lower the predicted days to cracking by no more than 10 percent, generally closer to 5 percent. There would still be good agreement between predicted and observed values for the 5-sack mix shown on Fig. 4. For the 7½-sack mix the prediction $C = 289$ days is plotted in Fig. 4. This would imply that galvanized steel reinforcement resists cracking of concrete longer than black steel reinforcement, a statement which apparently contradicts the experimental data. However, Spellman and Stratfull [*23*] state that "the time to the active potential of steel in concrete is mathematically related to the time to concrete cracking due to corrosion." They also state that "visual observations not only are of questionable accuracy, depending on the observer, but also are a more time-consuming and expensive procedure than the measuring of half-cell potentials." Accordingly, one can have considerable confidence in the predicted time to cracking of concrete due to corrosion of black steel.

The fact that only seven out of 20 black steel specimens of the 7½-sack-mix concrete had cracked in 622 days when prediction indicated that the average time-to-cracking would be 289 days raises questions about these specimens. Fortunately, Spellman and Stratfull have published data for an 8-sack mix with no admixture, both for moist cure and for steam cure. These data are presented in Table 4, and points for the moist cure are shown in Fig. 5. For the moist cured: the 7½-sack-mix galvanized cracked

TABLE 4—*Average days to active potential and concrete cracking* [23,24].

Steel	Cement Content		Cure	Observed Days to Active Potential	Days to Concrete Cracking	
	kg/m^3	$Sacks/yd^3$			Observed	Predicted
Black	279	5	moist	55	175	177
			steam	40	124	160
	418	7½	moist	155	622	289
			steam	155	622	289
	446[a]	8	moist	348.6, 339.0	390.8, 406.8	505.4, 494.7
			steam	212.4, 174.8	341.2, 296.8	352.9, 310.8
Galvanized	279	5	moist	40	315	279
			steam	40	257	279
	418	7½	moist	160	549	532
			steam	105	387	417

NOTE: 1 $kg/m^3 = 0.062\ lb/ft^3$.
[a] For slumps of 5.1 and 10.2 cm (2 and 4 in.) after Ref *23*.

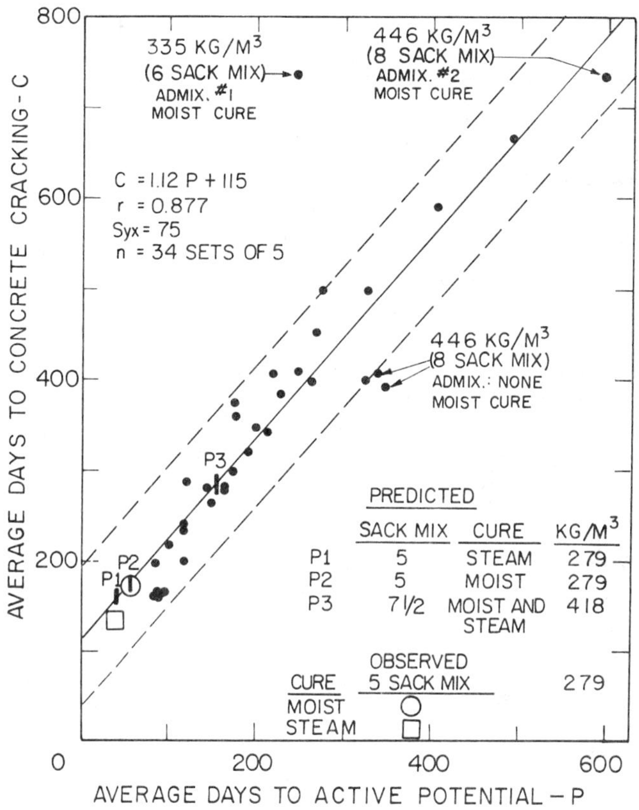

FIG. 5—*Days to concrete cracking versus days to active potential—black steel. Data and predictions of Table 4 are superposed on Fig. 3 of Spellman and Stratfull* [23]. *Laboratory corrosion test of steel in concrete. Active potential = −0.35 V saturated calomel electrode (SCE)* (*1 kg/m*^{3} *= 1.685 lb/yd*^{3}).

in 549 days (observed); the 7½-sack-mix black cracked in 289 days (predicted); and the 8-sack-mix black cracked in 390.8 and 406.8 days (observed). For the steam cured: the 7½-sack-mix galvanized cracked in 387 days (observed); the 7½-sack-mix black cracked in 289 days (predicted); and the 8-sack-mix black cracked in 341.2 and 298.8 days (observed). The two dates of cracking for the 8-sack mix refer to 5.08 and 10.16-cm (2 and 4 in.) slump concrete. The galvanized reinforcement does better than the black in these statistics, but it is not clear why the 1.8-to-1 improvement in performance found for the 5-sack mix is not maintained in the 7½-sack mix.

Tonini, Cook, and Cornet, in their discussions Ref *23*, point out that some of the authors' speculations are in error and misleading: speculation about reversal of polarity with zinc becoming cathodic to steel in a concentrated sodium chloride solution is one example. Another example is calling a solution "tap water" when the concrete in the atmospherically exposed zone of the specimen shows a 0.95-kg/m^3 (1.6 lb/yd^3) chloride ion. Sacramento city tap water contains 20 to 40 ppm of chloride ion, or 0.039-kg chloride ion per cubic metre (0.067-lb chloride ion per cubic yard) of water.

All of the foregoing discussion neglects the very important question: Does a "lollipop" prism specimen partially submerged in a saturated sodium chloride aqueous solution tell us anything about the performance of a reinforced concrete structure in atmospheric or even in marine installations? If one wants to build a brine concentration or a salt evaporation tank, the specimen design and exposure may be appropriate. Otherwise, it is questionable.

Bird and Strauss

Bird and Strauss [*39*] tested specimens of steel, high-purity tin, copper, cadmium, nickel, lead, and zinc. Each specimen was soldered at one end to a single-strand plastic-covered copper wire and the joint was insulated with epoxy tar. These specimens, 10.16 cm (4 in.) long by 0.63 cm (0.25 in.) in diameter, were embedded in 15.24 by 5.08 by 5.08-cm (6 by 2 by 2 in.) mortar blocks consisting of two parts by weight of clean river sand and one part by weight of portland cement. Similar specimens were embedded in mortar containing 1 percent sodium chloride by weight of cement. After curing for one week, specimens were partly immersed in fresh water, in 1 percent salt solution, and in seawater. Potentials were recorded against a saturated calomel electrode until steady values were obtained and rest potentials of the metals in seawater were determined. Other specimens were exposed two months in a high-humidity room.

Four-month salt spray tests were run on galvanized, nickel-plated, and cadmium-plated steel (not in concrete). Finally, zinc-dipped [0.305 kg/m^2 (1.0 oz/ft^2)] and cadmium-dipped [0.161 kg/m^2 (0.53 oz/ft^2)] steel rods were embedded centrally in concrete blocks with an aggregate/cement ratio of 6.5 and water/cement ratio of 0.75, cured seven days, then exposed to a salt spray, 3 percent salt, for three years, operated on alternate days with specimens allowed to dry in between.

Of the metallic coatings examined, both in the laboratory and under exposure conditions, cadmium provided the most satisfactory protection to reinforcing steel in concrete. Bird and Strauss concluded that a galvanized coating on reinforcing steel did postpone the onset of corrosion of steel, but would not delay its occurrence for a very long time.

Baker et al

Baker et al [*40*] investigated concrete beams 15.24 by 15.24 by 60.96 cm (6 by 6 by 24 in.) with four reinforcing bars in each, exposed at test lots at Wrightsville Beach, N.C. (The International Nickel Co.). Uncoated, hot-dip galvanized, and nickel-coated steel bars were used. Concrete with W/C ratios of 0.66 and 0.71 and cement content of 3.1 sacks/m^3 (5.3 sacks/yd^3) was used with thickness of cover at 1.27 and 3.2 cm (0.5 and 1.5 in.). Exposure conditions were (1) splash-and-spray zone 1.5 m (5 ft) above mean high tide, (2) tidal zone with alternate wet and dry conditions, and (3) elevated platform 24.3 m (80 ft) from mean high tide. The test ran for 11 years. The onset of corrosion was determined in terms of cracking and spalling of concrete, visual observation of the condition of reinforcing bars after removal from concrete specimens, and metallographic examination of coated reinforcing bars to determine loss of coating thickness.

Concrete with metal-coated reinforcing bars performed substantially better than concrete with uncoated bars, based on the degree of concrete cracking and spalling and on the degree of rusting of bars. The performance of galvanized and nickel-coated bars was essentially the same. Specimens exposed in the splash zone and elevated platform suffered the most corrosion damage, while those in the tidal zone showed no cracking.

Martin and Rauen

Martin and Rauen [*41*] ran tests to determine (1) changes experienced in the zinc layer on reinforcing steel in concrete and (2) whether galvanizing leads to improved protection against corrosion in carbonated concrete.

For Part 1 of their tests they used a 4 by 4 by 16-cm (1.6 by 1.6 by 6.4 in.) concrete prism reinforced with a 12-mm-diameter (0.48 in.) galvanized steel bar centrally located lengthwise in the prism. Both plain and deformed bars were used, with galvanized coatings 615 ± 123 g/m^2 (0.13 lb/ft^2). Test specimens were made of two different types of portland cement and two types of blast furnace slag cement with different amounts of natural chromate. There were three variations of zinc: a standard galvanized coating, a galvanized coating followed by galvannealing treatment, and a galvanized coating with a low content of aluminum to obtain a pure zinc coating. Concrete with water/cement ratios of 0.60 to 0.61 and cement contents of 350 and 370 kg/m^3 (6.3 and 6.6 sacks/yd^3) were used.

Reinforced concrete prisms were exposed at 20°C (68°F) at 40, 60, 80, and 100 percent relative humidity for periods of 7 days, 1 month, 6 months, 1 year, and 2 years. Zinc consumption was determined as a function of time. It was found that zinc was consumed mainly in the first 7 days, but at 100 percent relative humidity for 2 years' exposure there was additional zinc removal.

Similar specimens were exposed to alternate immersion and up to 64 cycles of alternating wet storage and dry storage. Again, zinc consumption was measured.

Another series of tests was run to determine the effects of carbonation. Here concrete prisms 4 by 4 by 60 cm (1.6 by 1.6 by 6.4 in.) were used, reinforced with 12-mm-diameter (0.48 in.) deformed bars. Test specimens were stored for one month in the laboratory, then carbonated in an air-conditioned room with a 3 percent CO_2 atmosphere. The significant finding is that, immediately after carbonation of the concrete, the ungalvanized bars began to rust, and by two years had firmly adhering rust crusts, but the galvanized bars showed no signs of rust after two years of exposure.

The analytic procedure used for determining zinc loss overestimated the amount of zinc removal because the authors assumed that the attack on zinc by the acetic acid reagent was negligible. Using these high estimates of zinc loss, for specimens with a thin [14 mm (0.56 in.)] cover of poor-quality mortar (W/C = 0.6), the investigators reported that 85 percent of the zinc coating remained after two years of exposure.

Nishi

Nishi [*42*] reported on an investigation dealing with the corrosion performance of several types of specimens, such as reinforced concrete beams and long concrete cylinders with axial reinforcing steel bars, and built-in gaps in the concrete (described as concrete disks "spitted" on a steel bar). Companion specimens reinforced with black and galvanized steel were included in this series of tests. Dimensions of the specimens are given in Table 5.

Several types of exposure were used: (1) in the laboratory—a salt spray fog with varying amounts of SO_2 fumes (from 100 to 1500 ppm) and exposure times of 20, 30, and 70 days; and (2) in the field—a seaside industrial exposure for 6 months and for 1 year.

Results of tests on beams showed that galvanized reinforcing bars, as compared with black bars, had no rusting for 20 and 30 days of exposures, but for 70 days of exposure had some rusting. Black bar specimens exhibited

TABLE 5—*Types, number, and dimensions of specimens* [42].

		Beams		Spitted Disks
Dimensions of specimen, mm		80 × 40 × 400		50 × 50 × 230
Dimensions of bar, mm; diameter and length		6φ × 440		9φ × 190
Treatment of steel		galvanized steel black steel		galvanized steel black steel
Cement		rapid-hardening portland		rapid-hardening portland
Crack or slit width, mm	Crack	0.2	0.5	. . .
	Slit	. . .		10
No. of specimens per exposure age		8 (4 sets of 2)		4

corrosion at 20 days. Corrosion of reinforcement was greater at cracks in the beams.

The spitted-disks-type specimens showed a rusting pattern similar to the cracked beam specimens, but the 70-day laboratory exposure tests showed decreasing difference between black and galvanized bar specimens. In the outdoor exposure tests, the degree of resistance to rusting of galvanized reinforcing bars was higher than for black bars in all specimens.

The results of Nishi's corrosion performance investigations can be summarized as follows:

(1) The durability of reinforced concrete exposed to sodium chloride solution or seawater will be improved by using galvanized reinforcing bars.

(2) Using galvanized steel, the durability of concrete members with cracks of about 0.3 mm (0.012 in.) will be equal to the durability of ordinary reinforced concrete with cracks of about 0.2 mm (0.008 in.).

Bond Characteristics

Bond between reinforcing steel and concrete is essential for reliable performance of reinforced concrete structures. Because corrosion can adversely affect this bond, various methods of protecting steel reinforcement have been sought. Two main types of test specimens have been used: pullout specimens, in which a single reinforcing steel bar embedded in concrete is pulled out of the block, and beam or beam-segment specimens, in which a reinforcing steel bar is embedded in the tension zone of the specimen and a load on the steel bar is generated, simulating beam bending. Usually, as the tension load increases, slip at the loaded and unloaded ends of the embedded bar is measured. The pullout test, although commonly used, does not simulate the embedment conditions of the bar in the tension zone of a concrete beam because the concrete in the pullout block is usually subjected to compression, whereas the concrete in a beam or slab, where bond characteristics are important, is in tension. Therefore, results of pullout tests are at best indicators of relative performance of bars with different surface conditions; however, to the extent that corrosion may also be dependent on state of stress in concrete, for example, cracked versus uncracked, the surface conditions of corroded steel in pullout specimens are not representative of corrosion in the tension zone in the beam.

Therefore, results of pullout tests should not be used as a measure of bond resistance of steel reinforcement in concrete, particularly when the specimens are exposed to aggressive environment prior to test.

Slater et al

Slater et al [*43*] published results of a study on relative bond resistance of reinforcing steel bars with 19 different surface conditions. Three of these conditions are of interest in this review: ungalvanized (UG), hot-dip galvanized (HDG), and electrogalvanized (EG). Pullout test specimens were used; the concrete block was a 15.24-cm-diameter (6 in.), 15.24-cm-long (6 in.) cylinder, with a 1.27-cm-diameter (½ in.) 60.9-cm-long (24 in.) steel bar embedded in the block. Two types of steel bar were used: a plain smooth bar with yield strength of 228 MPa (33 ksi), and an "old" deformed (now classified as "plain") bar with yield strength of 421 MPa (61 ksi). For the concrete block, a mortar mix with W/C ratio of 0.46 by weight, a cement/sand ratio of 1:2, and 0.635-cm (¼ in.) maximum-size sand were used. The compressive strength of the mortar, based on tests of moist-cured specimens at 28 days, was reported as 34 474 kPa (5000 psi).

All pullout specimens were moist cured for 28 days in damp sand, and subsequently exposed to four different environments prior to testing, as follows: (1) one day in dry air, (2) five months in dry air, (3) five months continuous immersion in simulated "seawater," and (4) five months of alternating drying and immersion in "seawater" using 12-h cycles. The results of these tests are summarized in Table 6. As seen from the table, the series of tests is not complete. Specimens with EG deformed bars were not tested, and specimens with HDG bars, both plain and deformed, were not subjected to alternating drying and immersion cyclic exposures. Nevertheless, some important results were observed:

1. HDG bars developed significant increase in bond strength during the five months following initial curing. Plain bars increased bond strength from 2758 to 3447 kPa (400 to 500 psi), 25 percent (in dry air), and from 2758 to 4137 kPa (400 to 600 psi), 50 percent (in continuous "seawater" immersion). "Deformed" bars increased bond strength from 3103 to 7584 kPa (450 to 1100 psi), 140 percent (in dry air), and from 3103 to 6550 kPa (450 to 950 psi), 110 percent (in continuous "seawater" immersion).

TABLE 6—*Average bond strength at failure after different exposures* [43].

Bar surface	Plain			Deformed		
Bar coating[a]	UG	HDG	EG	UG	HDG	EG
Exposure:	Bond Stress, ksi[c]					
Dry air—1 day	650	400	600	1100	450	. . .[b]
Dry air—5 mos.	650	500	580	1250	1100	. . .
"Seawater":						
continuous	750	600	650	1080	950	. . .
alternating	750	. . .	750	950	. . .	. . .

[a] Bar coatings: UG = ungalvanized; HDG = hot-dip galvanized; EG = electrogalvanized.
[b] No data.
[c] 1 ksi = 6.8948 MPa.

2. EG bars did not develop a consistent change in bond strength during the five months following initial bond strength. In "seawater" with continuous immersion there was an 8 percent increase in bond strength, and with alternating drying and immersion there was a 25 percent increase in bond strength.

3. UG plain bars developed no change in bond strength during five months' dry air exposure after initial curing. In "seawater" a 25 percent increase in bond strength was developed for both continuously immersed and alternately dry and immersed specimens. UG "deformed" bars in dry air developed a 1 percent increase in bond strength during five months. In "seawater," however, a 2 percent decrease was noted for continuously immersed specimens, and a 1 percent decrease for alternately dried and immersed specimens.

4. Generally, galvanized bar specimens *after five months' exposure* showed some reduction in bond strength compared with UG bars. The reduction in strength of plain bar specimens varied from 0 percent (EG bars exposed to alternating drying and immersion) to 23 percent (HDG bars exposed to dry air). The reduction for "deformed" bars was of the order of 12 percent.

5. HDG bars tested one day after a 28-day cure showed significantly lower bond strength than UG bars—about 3 and 5 percent lower values. EG bars tested at the same age showed only slightly lower bond strength than UG bars.

Even with limited data available in this study, three important conclusions emerged.

1. Development of bond between steel and concrete is age and environment dependent.
2. In some cases, the time required for developing a full bond between steel and concrete may be greater for galvanized bars than for ungalvanized.
3. The fully developed bond strength of galvanized and ungalvanized deformed bars is not significantly different for exposures of varying severity.

Lewis

Lewis [*44*] published results of a study on pullout specimens which were used to evaluate relative bond resistance of mild steel bars (probably plain undeformed bars, although the type of bars is not specifically identified in the paper). The effect of two different exposures on bond strength of reinforcing steel bars, ungalvanized and hot-dip galvanized, placed either near the top or the bottom of the specimen, was investigated.

Concrete prism specimens, 10.1 by 30.4 cm (4 by 12 in.) in cross section, 41.9 cm (16.5 in.) long, and reinforced with two 0.95-cm-diameter (3/8 in.) steel bars with 11.4-cm (4.5 in.) embedment and 1.27-cm (1/2 in.) cover at top and bottom, were exposed to "artificial marine" and "inland rural" environments. The zinc coating on the HDG bars was about 0.610 kg/m^2 (2 oz/ft^2) and the exposure periods were 12 and 24 months. Ungalvanized bars were degreased prior to casting; the concrete mix proportion and concrete strengths were not reported. Test results are summarized in Table 7.

The data reported by Lewis are also limited in scope and not fully reported (for example, concrete mix proportions and strength are not reported); therefore, the data are also difficult to interpret. Nevertheless, two important conclusions emerge:

1. For a given exposure, both top and bottom galvanized bars developed greater bond strength at failure than as-received ungalvanized bars under similar conditions.
2. Galvanized top bars exposed to "artificial marine" environment exhibited approximately double the bond strength of the ungalvanized top bars.

Bresler and Cornet

Bresler and Cornet [*11*] reported a study on bond of galvanized and ungalvanized steel bars in which beam specimens were used and bond of top and bottom bars was investigated. This series of tests differs from other studies in two respects: beam specimens were used in lieu of pullout specimens, and slip between steel and concrete at various levels of bond stress was reported. In the summary, only bond strengths for different conditions are reported.

TABLE 7—*Average bond strength at failure after different exposures* [44].

Bar position	Bottom		Top	
Bar coating[a]	U	HDG	U	HDG
Exposure:	Stress, ksi			
Inland rural				
12 months	745	800 (y)[b]	700	800 (y)
24 months	635	750	680	800 (y)
Artificial marine				
12 months	650	800 (y)	180	430
24 months	660	800 (y)	270	460

[a] Bar coatings: U = as received; HDG = hot-dip galvanized.
[b] (y) = yield reached in steel reinforcement prior to reaching bond capacity.
[c] 1 ksi = 6.8948 MPa.

Beam specimens conformed to American Concrete Institute Standard 208-58. The beam specimen cross section was 20.3 by 45.7 cm (8 by 18 in.) and the span 182.8 cm (72 in.). The beams were tested in flexure using equal concentrated symmetrical loads spaced at 121.9 cm (48 in.). Bond stresses were determined in the region of constant bending moment. Each beam cast erect was reinforced with one top and one bottom bar, 1.27 and 2.2 cm (½ and ⅞ in.) diameter, respectively. For beams cast inverted the sizes of reinforcing bars were reversed, so that bond stresses were always determined for the 2.2-cm (⅞ in.) bar. Plain and deformed steel bars with yield strengths of about 276 MPa (40 ksi) were used. Galvanized coating provided zinc in an amount about 0.915 kg/m^2 (3 oz/ft^2) of bar surface. Both galvanized and ungalvanized bars were lightly corroded prior to casting of the beams. The following procedure was used:

1. Galvanized bars: 3 days in an aerated water bath at 16 to 27°C (60 to 80°F), followed by 59 days in a "fog room" at 21°C (70°F) and 100 percent relative humidity.
2. Black bars: 6 days in an aerated water bath at 16 to 27°C (60 to 80°F), followed by 56 days in a "fog room."

The average compressive strength of the concrete was 28 958 kPa (4200 psi). Beams were moist cured for 14 days and tested after about 14 days of drying. Test results are summarized in Table 8. While these tests did not reflect the influence of different environmental exposures, two important conclusions emerged:

1. For plain bars, the bond strength of galvanized reinforcement was 30 to 50 percent greater than for similar ungalvanized reinforcement.
2. For deformed bars, the bond strength of galvanized bars was the same as that for similar ungalvanized bars.

Hofsoy and Gukild

Hofsoy and Gukild [*45*] published results of an investigation of the bond of HDG reinforcement in concrete. Pullout test specimens were used,

TABLE 8—*Average bond strength at failure* [11].

Bar surface	Plain				Deformed			
Bar position	Bottom		Top		Bottom		Top	
Bar coating[a]	UG	G	UG	G	UG	G	UG	G
Bond strength, ksi[b]	360	480	110	170	850	850	800	800

[a] UG = ungalvanized; G = galvanized.
[b] 1 ksi = 6.8948 MPa.

consisting of 10-cm (4 in.) concrete cubes with 8-mm-diameter (0.31 in.) plain bars and 10-mm-diameter (0.39 in.) deformed bars with 9- and 8-cm (3.6 and 3.2 in.) embedment lengths, respectively. The specimens with deformed bars had ungalvanized spiral reinforcement to prevent splitting of concrete cubes.

Plain bars were of cold-drawn steel with a yield strength of about 462 MPa (67 ksi), and with a smoother surface than for bars usually used as reinforcement for concrete. Deformed bars were hot-rolled steel with an average yield strength of 427 MPa (62 ksi).

Hot-dip galvanized bars were processed at a steel mill with about 0.305 kg/m^2 (1 oz/ft^2) for the plain bars and 1.372 kg/m^2 (4.5 oz/ft^2) for deformed bars. The age of zinc coating was about 1 to 2 months for plain bars, 36 months for deformed bars. All bars were stored at about 20°C (68°F) temperature and a relative humidity of 46 to 55 percent. The ungalvanized deformed bars were slightly rusted during storage, but this was "cleaned off" before embedment. Some galvanized bars were chromatized by dipping in a saturated solution of $K_2Cr_2O_7$ in a diluted sulphuric acid.

Three brands of cement were used, two with low chromate and one (No. 3) with relatively high chromate content. River sand and crushed stone with maximum size of 1.58 cm (5/8 in.) were used as aggregates. The proportions of the mix were 1:2.5:2.5 with a W/C ratio of 5.5. In specimens where unchromatized galvanized bars were used, a chromate admixture ($K_2Cr_2O_7$) was used. Test results are summarized in Table 9.

Data reported by Hofsoy and Gukild [*45*] are unfortunately based on tests of moist-cured specimens at an age of 7 days. As noted earlier (see Slater et al [*43*] study), bond strength in some cases increases significantly with age, and the results with different cements and passivation procedures

TABLE 9—*Average bond strength at failure with different cements* [45].

Bar surface	Plain				Deformed			
Bar coating[a]	U	HDG			U	HDG		
Passivation[b]	N	N	CRA	CRD	N	N	CRA	CRD
Cement[c]	Ratios of Bond to Compressive Strength							
A	0.09	0.06	0.11	...[d]	0.40	0.20	0.38	0.45
B	...	...	...	...	0.33	0.25	0.28	0.38
C	0.09	0.12	...	...	0.42	0.38	0.38	...

[a] Bar coatings: U = ungalvanized; HDG = hot-dip galvanized.

[b] Passivation: N = none; CRA = chromate admixture to concrete; CRD = chromatized after galvanizing.

[c] Cements: A = low chromate; B = low chromate; C = high chromate.

[d] No data.

could be influenced differently by aging. Also, the data on plain bars were obtained using smooth cold-rolled steel bars which are not representative of reinforced concrete practice. Therefore, results of these tests cannot be used directly in establishing bond strength values for reinforced concrete under realistic conditions of construction.

Nevertheless, the Hofsoy and Gukild tests demonstrated most clearly the importance of passivation of galvanized reinforcement in concrete. Where such passivation was obtained by use of cement with high chromate content, substantially the same bond strength was obtained for galvanized and ungalvanized deformed bars. Where passivation was obtained by chromatizing the bars after galvanizing, the bond strengths for galvanized deformed bars were greater than those for ungalvanized similar bars. When passivation was obtained by addition of chromate to the concrete mix, some scatter was observed: in one case, the bond was essentially the same (Cement No. 1); in another (Cement No. 2), about 20 percent decrease was observed. Since the Hofsoy and Gukild tests were made at the age of 7 days and since bond resistance is age-dependent, as demonstrated by Slater et al in 1920 [*43*], no general conclusions can be drawn from the data described in the foregoing.

Nishi

Nishi [*42*] also investigated the effect of galvanizing on the bond strength of reinforcing steel bars having different shapes and patterns of deformations. Special deformed bars with 10 different shapes and spacing of lugs were machined from plain round steel bars, and pullout specimens were made using both black and galvanized steel. Although the test results varied somewhat with type of deformation, they indicated that the bond of galvanized deformed bars, meeting normal standards for shape and spacing of lugs, was at least as good as that of black bars.

Mechanical Properties

Ductility

The ductility of reinforcing steel is important to ensure ductile behavior of reinforced concrete and to prevent brittle failure in steel details. It is important, therefore, to investigate whether galvanizing may adversely affect the ductility of reinforcing steel.

Lewis and Booker [*46*] reported a program which was designed to determine the effect of the galvanizing process on the ductility of steel bar anchors and inserts after they had been subjected to different fabrication procedures. The specimens were made of 0.95-cm-diameter (3/8 in.) steel bars: hot-rolled rounds [ASTM Specification for Structural Steel (A 36-75)] and deformed reinforcement of intermediate grade. Several treatments prior to galvanizing were investigated, and bond tests were conducted on straight bars, straightening tests on prebent bars, and bend tests on welded bars. Three pin diameters were used in bend tests: one, two, and six bar diameters. Bars, black and galvanized, bent on a "one-diameter" pin, showed considerable reduction in ductility, and one third of the specimens failed. Bars bent on two-diameter and six-diameter pins showed good ductility. The results demonstrated conclusively that when proper procedures are followed galvanizing has no effect on the ductility of black steel.

Strength and Ductility

Bresler and Cornet [*12*] carried out a pilot study on the effect of galvanizing on mechanical properties of steel reinforcing bars. Standard tension and bend tests were performed on No. 7 deformed steel bars of four ASTM grades (based on 1967 Standards) as follows: A-15 (intermediate); and A-431, A-432, and A-432 (special grade which satisfied A-431 yield strength requirement while retaining the ductility of A-432 grade). One black and one galvanized bar specimen were obtained from the same bar and duplicates from two bars from the same heat. Test specimens were galvanized by a commercial hot-dip galvanizer; all bars had 0.152 to 0.203 mm (6 to 8 mils) of zinc coating, averaging somewhat better than 0.915 kg/m^2 (3 oz/ft^2). Test results are summarized in Table 10, where each result represents an average of two test specimens.

Studies of the effect of galvanizing on the strength and ductility of reinforcing steel have been also carried out by Martin and Rauen [*41*] and

TABLE 10—*Mechanical properties of reinforcing steel bars* [12].

Grade	A-15 Intermediate		A-431		A-432		A-432[a]	
Coating[b]	B	G	B	G	B	G	B	G
Elastic modulus, 10^3 ksi	25.3	27.8	28.1	30.5	26	26	28.2	29.9
Yield strength, ksi	46.3	48.3	99.3	102.8	67.3	69.0	76.1	76.3
Tensile strength, ksi	71.8	73.5	140	137.5	92.5	92.7	108.3	108.8
Elongation, 8 in.-%	28.5	20.0	[c]	[d]	18.0	17.0	11.6	13.0
Bend test[f]	OK	OK	OK	[e]	OK	OK	OK	OK

[a] Special grade: A-431 yield strength with A-432 ductility.
[b] B = black; G = galvanized.
[c] One specimen had 9.1% elongation, passing the requirement of 7%; another specimen had elongation of 4.37, failing to meet the specified value.
[d] The specimen elongations were 6.48 and 5.24%, both failing to meet the 7% requirement.
[e] One specimen passed bend test requirements and the other failed.
[f] Bend tests used 3½-in.-dia pin for A-15 steel and 4.5-in.-dia pin for other grades.
NOTE: 1 ksi = 6.8948 MPa; 1 in. = 2.54 cm.

by Nishi [42]. Both investigations demonstrated that galvanizing had no negative influence on the tensile strength or on the ultimate strain of the steel, and these results are fully consistent with the conclusions of Bresler and Cornet [12].

Fatigue Strength

Nishi [42] conducted fatigue tests on 60 reinforced concrete 20-cm-high (8 in.) and 160-cm-long (64 in.) rectangular beams with galvanized or black steel bar reinforcement. Twenty-four beams in cracked condition were tested after exposure to sodium chloride solution for 6 or 12 months. The results showed that galvanized bars provided improved fatigue resistance.

Corrosion fatigue of prestressing wires, black and galvanized, was studied in a rotary bending tester in a saturated calcium hydroxide solution containing 2 percent calcium chloride. In this solution, black wires suffered from pitting and failed after 10^5 cycles under a stress of 40 kg/mm^2 (56 774 psi). Galvanized wires were nearly passive in the solution and were not significantly corroded, attaining 10^7 cycles under a stress of 48 kg/mm^2 (68 272 psi).

Fatigue resistance of six different types of deformed bars, ranging in diameter from 13 to 32 mm (0.52 to 1.28 in.), both galvanized and black, was investigated. Prior to testing, the bars were exposed at a seaside industrial area for a 12-month period. The allowable tensile stresses (kg/cm^2) for a 32-mm-diameter (0.52 in.) steel bar with diagonal lugs under three types of alternating stress are given in Table 11.

Field Performance

Duffaut et al

Duffaut et al [47] reported tests conducted in the Rance Estuary from 1959 to 1971. Specimens were concrete prisms 0.20 by 0.20 by 1.20 m

TABLE 11—*Allowable tensile stresses for 32-mm-dia bar* [42].

Coating	Stress Range, kg/cm^2		
		Tension	
	Full Stress Reversal	Zero to Maximum	1:2
Black	880	1500	2300
Galvanized	1000	1700	2600

NOTE: 1 kg/cm^2 = 98.066 kPa.

(0.65 by 0.65 by 3.93 ft) with one reinforcing bar, and 0.40 by 0.40 by 1.20 m (1.30 by 1.30 by 3.93 ft) with four reinforcing bars, and different thickness of cover over each bar. The reinforcing bars were 10 mm (0.4 in.) in diameter. Two mixes of concrete were used, one "good," the other "bad," with water/cement ratios of 0.52 and 0.96. Some specimens were cracked by bending, others were not cracked. Some specimens had black steel, some had galvanized steel, some were protected by cathodic protection and others were not. Specimens were exposed to seawater at five levels in the tidal zone from continual immersion at 0.0 m to spray exposure at +13.50-m (44.28 ft) elevation.

Performance was determined by measurement of potential, current, and resistance, and by visual examination of specimens for distress and classification of damage from 0 (no corrosion damage) to 4 (severe corrosion damage).

With both good- and bad-quality concrete, galvanizing effectively reduced the degree of corrosion of steel reinforcement. The effectiveness of galvanizing was less at the upper level, in the spray zone, with bad-quality concrete, but performance of specimens with galvanized bars was better than with black bars. Galvanized steel bars in bad concrete without cathodic protection performed almost as well as black bars in bad concrete with cathodic protection. Best performance was obtained with either good or bad concrete with galvanized reinforcement plus cathodic protection.

Stark and Perenchio

Stark and Perenchio [*27*] reported on elevated outdoor test slabs exposed at the Portland Cement Association, Skokie, Ill., to obtain a comparison of performance of black and galvanized reinforcing steel in simulated deck slabs subjected to chloride deicer salts. Slabs were 1.21 m (4 ft) wide by 1.52 m (5 ft) long by 15.24 cm (6 in.) deep, each containing two mats of reinforcing steel. Each slab was treated with calcium chloride deicer salts about 15 times each winter at the rate of 0.488 kg/m^2 (0.1 lb/ft^2). The slabs carried slotted dikes which permitted snow to collect but allowed deicer solution to drain, thus simulating bridge deck conditions.

The concrete mix had a cement content of 4.56 $bags/m^3$ (6 $bags/yd^3$), a W/C ratio of 0.43 to 0.44, and an air content of 5.5 to 6.2 percent.

There were four slabs: two had black and two had galvanized reinforcing steel. One of each of these had 1.27-cm (½ in.) concrete cover over the reinforcing steel; the other had 3.81-cm (1½ in.) concrete cover.

This was a highly sophisticated study, involving electric potential measurements, mapping of surface defects, sounding for surface delaminations, chloride analyses and pH measurements, petrographic examination of concrete cores, and metallographic examination of galvanized coatings.

The slab with black steel reinforcement and 1.27 cm (½ in.) of cover exhibited rust stains two years after construction, and after the third year of exposure there was rusting, cracking, and spalling at the top (wearing) surface. In contrast, the slab which contained galvanized steel reinforcement with 1.27-cm (½ in.) cover showed no evidence of distress after six years of exposure. Analyses of chloride ion in the concrete showed a concentration of chloride at the level of the reinforcement which would be corrosive to steel, taken to be 0.65 to 0.77 kg of chloride ion per cubic metre (1.1 to 1.3 lb of chloride ion per cubic yard) of concrete.

In the slabs with 3.81-cm (1½ in.) cover, the chloride levels indicated that a corrosive concentration has been attained only on the slab with black steel, and neither of the slabs showed distress after six years' exposure.

The remainder of the report dealt with investigations of six bridge decks: Boca Chica Bridge, Fla.; Seven Mile Bridge, Fla.; Longbird Bridge, Bermuda; Flatts Bridge, Bermuda; I35 Bridge, Iowa; and Manicougan Bridge, Que. This investigation showed that in several of these structures, reinforcing steel is exposed to an environment corrosive to untreated steel, where a corrosive environment is defined by chloride concentration in excess of 0.65 to 0.77 kg of chloride ion per cubic metre (1.1 to 1.3 lb of chloride per cubic yard): Longbird (21 years' exposure), Boca Chica (3 years), and Seven Mile (3 years). Where untreated steel was present, there was evidence of corrosion, and in several cases associated distress in the concrete. Where galvanized steel was used, there was no evidence of corrosion or impaired performance of the concrete.

Okamura and Hisamatsu

Okamura and Hisamatsu [*48*] investigated corrosion of high-strength deformed black steel bars and galvanized bars with zinc coating of 0.6 kg/m^2 (2 oz/ft^2) in concrete beams 170 by 200 by 1600 mm (6.8 by 8.0 by 64 in.). Concrete was made with high early portland cement with a 28-day compressive strength of about 40 MPa (5.8 ksi). The concrete water/cement ratio was 0.55. Slump was 40 mm (1.6 in.) and air content was 3 percent.

Beams were precracked by bending. They were then exposed outdoors at the University of Tokyo, and sprayed with 3 percent chloride solution twice daily except Sundays and holidays.

Bending fatigue tests were conducted after exposure of 6 to 12 months. Corrosion of reinforcement at cracks caused a decrease in fatigue strength. The longer the duration of exposure, the greater the reduction in strength. For black steel bars before exposure the 2 million cycle fatigue strength was 267 MPa (39 ksi), while for galvanized bars fatigue strength was 250 MPa (36.5 ksi); this difference in strengths was found negligible compared with the effect of bar deformations on fatigue. For black steel bars, fatigue strength was reduced to 200 MPa (29 ksi) after 6 months of exposure and to 167 MPa (24 ksi) after 1 year of exposure. For galvanized steel bars, the fatigue strength was 200 MPa (29 ksi) after 8 months of exposure, and 195 MPa (28 ksi) after 1 year of exposure.

The greater the width of cracks in the beams, the greater the decline in fatigue strength after exposure to the aggressive environment, but the beams with 0.3-mm-wide (0.012 in.) cracks and galvanized reinforcement had durability equal to that of beams with 0.2-mm-wide (0.008 in.) cracks and black steel reinforcement.

Thus the use of galvanized reinforcement in concrete beams gave the same fatigue strength as did black reinforcement in the absence of corrosion, and improved performance in a corrosive exposure.

Stark

Stark [*49*] reported a field investigation of four structures with seven specimens representing ages 7 to 23 years with a variety of exposures of galvanized reinforcement in concrete. He concluded that galvanized reinforcing bars are performing satisfactorily in normal-quality concrete with chloride concentrations well above that needed to induce corrosion of untreated steels. Table 12 presents his findings.

Concluding Remarks

Many difficult problems arise in devising testing procedures which relate to measuring corrosion performance of steel reinforcement in concrete structures. A listing of some of these problems is given here, followed by a brief discussion of each:

1. Isolating effects of corrosion from other processes of deterioration.
2. Establishing similitude or scaling relationships for size, exposure acceleration, etc.
3. Interpretation of test results, particularly with respect to statistical validity and reliability.
4. Establishing performance requirements and criteria for evaluating fulfillment of these requirements.
5. Establishing test objectives and design of tests to meet these objectives.

Corrosion of steel in concrete is only one of several possible processes of degradation and it is sometimes a consequence of other physicochemical processes (such as freeze-thaw effects or expansive aggregates) rather than a primary cause of degradation. Isolating the effects of corrosion from other factors may be difficult.

Corrosion depends on many variables which are interdependent. Because of the complex nature of the physical and electrochemical factors in the corrosion process and because of their coupling with each other, it is

TABLE 12—*Performance of field structures* [49].

Specimen	Age, years	Pounds of Chloride per Cubic Yard of Concrete at Level of Steel	Average Corrosion Layer Thickness, mils	% Coating Remaining
St. George (SG 17)	7	5.0	0.1	98
Bermuda				
Yacht Club (BYC 3)	8	6.1	none	100
Hamilton (H 22)	10	3.2	0.2	95
Hamilton (H 26)	10	6.0	0.3	96
St. George (SG 10)	10	7.7	0.2	99
St. George (SG 9)	12	10.7	0.5	92
Longbird (LB 20)	23	7.3	0.2	98

NOTE: 1 lb/yd^3 = 0.5932 kg/m^3; 1 mil = 0.0254 mm.

difficult to establish similitude or scaling relationships which would account for variations in thickness of cover, sustained and repeated stresses, spacing and width of cracks, maximum size of aggregate, size of reinforcing steel, temperature and humidity levels, freezing and thawing, salt concentration, oxygen concentration, stray and galvanic currents, and imposed electrochemical potentials. Large variations in field performance and in test results have been observed in what are purported to be similar structures or test specimens and under what were purported to be similar conditions. Therefore, planning field and laboratory tests for statistical validity and reliability is particularly important.

Also, in planning of field and laboratory tests, it is essential to define performance requirements. Four primary requirements are proposed:

1. structural integrity,
2. functional integrity,
3. durability, and
4. aesthetic appearance.

For each of these requirements a set of criteria which indicate compliance with the requirement must be defined. For example, one criterion for structural integrity deals with loss of cross section of reinforcement which relates to strength and stability of the structure; another criterion deals with potential loss of bond. Functional integrity criteria deal with problems of excessive deformation, degradation of concrete cover, leakage, and similar criteria of functional performance. Durability criteria deal with frequency and cost of maintenance to ensure structural and functional integrity. Aesthetic appearance criteria deal with both objective and subjective requirements of aesthetic quality.

Some of the criteria for corrosion performance requirements are summarized in Table 13.

When test procedures are planned in a relationship to well-defined criteria and performance requirements, the results of such tests can be interpreted with greater clarity and can lead to significant conclusions.

Often test results have not been planned in accordance with the preceding considerations in mind. Many of the tests reported in the literature have specific and narrow objectives. Interpretation of such tests with respect to their general validity and significance may be questionable.

A comprehensive review of literature, from which selected summaries have been included in this paper, leads to the following conclusions.

1. There are no standardized or widely accepted test procedures to evaluate corrosion performance of reinforcing or prestressing steel in concrete exposed to aggressive environment.

2. There is no general agreement on what constitutes a suitable environment for accelerated exposure tests.

3. Many tests have been reported without reference to performance requirements or criteria. Interpretation of such tests is often controversial.

4. There are numerous reports on corrosion performance of galvanized steel in concrete. These reports generally agree, sometimes explicitly and sometimes implicitly, that zinc coating furnishes cathodic protection to the steel reinforcement; there is less pitting and less intense and less extensive corrosion of the steel until substantial quantities of galvanized coating have been consumed. Also, the following observations have been substantiated by numerous investigators:

(*a*) Galvanized steel in concrete tolerates a higher chloride concentration than black steel before corrosion starts.

(*b*) Cracking of concrete cover in saline exposures may occur later with galvanized reinforcement than with black reinforcement, but this effect may be less pronounced for 7- or 8-sack mix than for 5- or 6-sack mix.

(*c*) Where reinforcement is exposed to coastal or marine atmospheres for some time before it is covered with concrete, galvanizing furnishes

TABLE 13—*Corrosion performance criteria.*

1. Structural Integrity	2. Functional Integrity	3. Durability	4. Aesthetic Appearance
(*a*) Reduction in steel section	(*a*) excessive deformation	(*a*) maintenance-free time	(*a*) red stains
(*b*) loss of bond	(*b*) degradation of concrete cover	(*b*) maintenance cost	(*b*) cracking (*c*) spalling

useful protection, prevents formation of loose rust, and gives better bond to concrete. The improved performance can be substantial, particularly for thin-shell and ferrocement construction.

5. Performance of galvanized steel in concrete depends in part on the alkalinity of the cement and on the presence of hexavalent chromium ion in the concrete/metal interface. Therefore, for general validity of tests on galvanized steel in concrete, the reinforcing steel should be treated with chromate, or a CrO_3-type admixture should be added to the concrete. Cement with controlled alkalinity should be used.

Acknowledgments

The authors gratefully acknowledge a series of research grants by the International Lead Zinc Research Organization, Inc. which made this study possible. The assistance of R. DiGennaro, assistant professor of mechanical engineering technology, Cogswell College, San Francisco, Calif., and Dr. Y. Gau, engineer, Union Carbide Corp., Bound Brook, N.J., is also gratefully acknowledged.

References

[1] Cornet, I., *Materials Protection,* Vol. 6, No. 3, March 1967, pp. 56–58.

[2] Cornet, I. in *Proceedings,* Third International Congress on Marine Corrosion and Fouling, Gaithersburg, Md., Oct. 1972, pp. 215–225.

[3] Dempsey, J. G. in *Proceedings,* American Concrete Institute, Vol. 48, No. 12, Oct. 1951, p. 157.

[4] Muller, P. P., *Concrete Research* (England), Vol. 6, No. 16, June 1954, p. 37.

[5] Pletta, D. H., Massie, E. F., and Robin, H. S. in *Proceedings,* American Concrete Institute, Vol. 46, 1952, p. 513.

[6] Vollmer, H. D. in *Proceedings,* 23rd Annual Meeting, Highway Research Board, Vol. 23, No. 42, 1943, p. 296.

[7] Roberts, M. H., *Concrete Research* (England), Vol. 14, No. 42, Nov. 1962, p. 143.

[8] Tomek, J. and Vaurin, F., *Zement-Kalk-Gips* (West Germany), Vol. 14, No. 3, March 1961, pp. 108–112.

[9] Arber, M. G. and Vivian, H. E., *Australian Journal of Applied Science,* Vol. 12, No. 3, 1961, pp. 339–347.

[10] Monfore, G. E. and Verbeck, G. J. in *Proceedings,* American Concrete Institute, Vol. 57, Nov. 1960, p. 491.

[11] Bresler, B. and Cornet, I. in *Proceedings,* Seventh Congress of the International Association of Bridge and Structural Engineers, Rio de Janeiro, Brazil, Aug. 1964.

[12] Bresler, B. and Cornet, I., "Mechanical Properties of Galvanized Steel Reinforcing Bars," unpublished, 1968.

[13] Lea, F. M. and Watkins, D. M., "The Durability of Reinforced Concrete in Sea Water," National Building Studies Research Paper No. 30, Department of Scientific and Industrial Research, London, U.K., Her Majesty's Stationery Office, 1960.

[14] Halstead, S. and Woodworth, L. A., *Transactions,* South African Institute of Civil Engineers, April 1955, pp. 115–134.

[15] Lewis, D. A. and Copenhagen, W. J., *Corrosion,* Vol. 15, July 1959, pp. 382–388; and *South African Industrial Chemist,* Vol. 11, No. 10, Oct. 1957, pp. 207–219.

[16] Griffin, D. F., "Corrosion of Mild Steel in Concrete," U.S. Naval Civil Engineering Laboratory, Technical Report R-306 Supplement, Aug. 1965.
[17] Griffin, D. F., "Corrosion of Reinforced Concrete in Marine Environments," Presented at the San Francisco Regional Meeting of the National Association of Corrosion Engineers, Oct. 1966; and Technical Note R-306 Supplement, Naval Civil Engineering Laboratory, Aug. 1965.
[18] Griffin, D. F., "Effectiveness of Zinc Coating on Reinforcing Steel in Concrete Exposed to a Marine Environment," Technical Note N-1032, Naval Civil Engineering Laboratory, July 1969.
[19] Griffin, D. F. and Henry, R. L., "The Effect of Salt in Concrete on Compressive Strength, Water Vapor Transmission, and Corrosion of Reinforcing Steel," Fourth Pacific Area Meeting, American Society for Testing and Materials, ASTM Paper No. 832, Oct. 1962.
[20] Blenkinsop, J. C., *Concrete Research* (England) Vol. 15, No. 43, March 1963, p. 33.
[21] Veits, R. I., *Stroitel'naya Promyshlennost*, No. 9, 1954.
[22] Hausmann, D. A., *A. P. Engineering Topics*, American Pipe and Construction Co., Dec. 1962.
[23] Spellman, D. L. and Stratfull, R. F., "Laboratory Corrosion Test of Steel in Concrete," Materials and Research Department, California Division of Highways, Research Department M&R 635116-3, Sept. 1968.
[24] Hill, G. A., Spellman, D. L., and Stratfull, R. F., "Laboratory Corrosion Test of Galvanized Steel in Concrete," Transportation Research Record No. 604, Transportation Research Board, Washington, D.C., 1976.
[25] Clear, K. C. and Hay, R. E., "Time to Corrosion of Reinforcing Steel in Concrete Slabs, Vol. 1, Effect of Mix Design and Construction Parameters," Report No. FHWA-RD-73-32 (Interim Report), Federal Highway Administration, April 1973.
[26] Clear, K. C., "Time to Corrosion of Reinforcing Steel in Concrete Slabs, Vol. 3, Performance after 830 Daily Salt Applications," Report No. FHWA-RD-76-70 (Interim Report), Federal Highway Administration, April 1976.
[27] Stark, D. and Perenchio, W., "The Performance of Galvanized Reinforcement in Concrete Bridge Decks," Final Report, Project No. 2E-206, Construction Technology Laboratories, Skokie, Ill., July 1974–Oct. 1975.
[28] Roetheli, B. E., Cox, G. L., and Littreal, W. B., *Metals and Alloys*, Vol. 3, March 1932, p. 73–76.
[29] Rehm, G. and Lämmke, A., *Deutscher Ausschuss für Stahlbeton*, Heft 242, Berlin, Germany, 1974, pp. 45–60.
[30] Duval, R. and Arliguie, G., *Memoires Scientifiques de la Revue de Metallurgie*, Vol. 71, No. 11, 1974.
[31] Bird, C. E., *Corrosion Prevention and Control*, July 1964, pp. 17–21.
[32] Cornet, I. and Bresler, B., *Materials Protection*, Vol. 4, No. 11, Nov. 1965, pp. 35–37.
[33] Cornet, I. and Bresler, B., *Materials Protection*, Vol. 5, No. 4, April 1966, pp. 69–72.
[34] Cornet, I., Bresler, B., and Ishikawa, T., *Materials Protection*, Vol. 7, No. 3, March 1968, pp. 45–47.
[35] Ishikawa, T., Cornet, I., and Bresler, B., in *Proceedings*, Fourth International Congress on Metallic Corrosion, Amsterdam, The Netherlands, 1972, pp. 556–559.
[36] Lorman, W. R., "Concrete Cover in Thin Wall Reinforced Concrete Floating Piers," Technical Note N-1447, Naval Civil Engineering Laboratory, July 1976.
[37] Sopler, B., "Corrosion of Reinforcement in Concrete," Cement and Concrete Research Institute, The Norwegian Institute of Technology, University of Trondheim, FCB-Report 73-4, 1973.
[38] Bernhardt, C. J. and Sopler, B., *Nordisk Betong*, Vol. 2, 1974.
[39] Bird, C. E. and Strauss, F. J., *Materials Protection*, July 1967, pp. 49–52.
[40] Baker, E. A., Money, K. L., and Sanborn, C. B. in *Chloride Corrosion of Steel in Concrete, ASTM STP 629*, American Society for Testing and Materials, 1976, pp. 30–50.
[41] Martin, H. and Rauen, A., *Deutscher Ausschuss für Stahlbeton*. Heft 242, Berlin, Germany, 1974, pp. 61–77.

[*42*] Nishi, T. (project manager), "Investigations on Mechanical Behavior of Galvanized Steel Reinforcement in Concrete," Japan Testing Center for Construction Materials, International Lead Zinc Research Organization Project No. ZE-170, Final Report, No. 5, 1 Jan. 1971 to 30 June 1974.

[*43*] Slater, W. A., Richart, F. E., and Scofield, G. G., Technical Paper No. 173, National Bureau of Standards, 1920.

[*44*] Lewis, D. A. in *Proceedings,* First International Congress on Metallic Corrosion, London, England, April 1961, pp. 547-555.

[*45*] Hofsoy, A. and Gukild, I., *Journal of the American Concrete Institute,* March 1969, pp. 174-184.

[*46*] Lewis, G. P. and Booker, P. P., "The Effect of Galvanizing on the Ductility of Steel Reinforcing Rods and Connections," Cominco Products Research Centre, Ont., Canada, Dec. 1968.

[*47*] Duffaut, P., Duhoux, L., and Heuze, B. in *Annales de d'Institut Technique du Batiment et des Travaux Publics,* Supplement No. 305, May 1973.

[*48*] Okamura, H. and Hisamatsu, Y., *Materials Performance,* July 1976, pp. 43-47.

[*49*] Stark, D., "Galvanized Reinforcement in Concrete Containing Chlorides," Final Report, Project No. 2E-247, Construction Technology Laboratories, Skokie, Ill., April 1976-April 1978.

[*50*] Cornet, I., Williamson, R. B., Bresler, B., Nagarajan, S., and Christensen, K. A. in *Proceedings,* Symposium on Concrete Sea Structures, Tbilisi, Georgia, U.S.S.R., Sept. 1972; Federation International de la Precontrainte, April 1973.

[*51*] Sarapin, I. G., *Promyshlennoe Stroitelistvo,* Vol. 36, 1958, p. 12.

Summary

John Slater of Packer Engineering Associates, Inc., delivered the keynote address, "Corrosion of Steel in Concrete—Knowledge and Needs," which set the stage for the papers following but which itself is not included in this volume. He pointed out that estimates by the Federal Highway Administration (FHWA) and the Salt Institute of the cost of salt-induced damage clearly show the side effects of using salt to maintain clear roads in winter, noting that these figures do not include potential losses due to transportation obstructions and delays. Although the FHWA has estimated the cost of protecting or rehabilitating highway structures only in the United States, chloride-induced corrosion is, in fact, a global problem, with particular impact in Canada and Europe. Approaches to preventing this corrosion deterioration range from surface-sealing the concrete to coating the steel (epoxy or galvanized) to use of cathodic protection or inhibitors. As Slater noted, highway structures are not alone in showing deterioration by chloride-induced corrosion of reinforcing steel. Vertical and lower surfaces of concrete buildings, wharves, and pilings need to be considered for protection. He suggested that before the problem can be considered truly under control, more information is needed regarding the interaction between moisture and oxygen in concrete: how much there is, and how it changes with time and temperature. Further definition of the rate of corrosion needs to be undertaken, and the effect of moisture and oxygen levels on this rate needs more study.

Carl Locke of the University of Oklahoma discussed "Electrochemistry of Salt-Contaminated Concrete." He showed that by using sodium chloride admixed in concrete with imbedded reinforcing steel, a critical sodium chloride concentration (0.1 to 0.2 percent based on total weight of concrete) increased the rate significantly, in agreement with other concentration threshold theories. Polarization resistance was used to ascertain corrosion rates, and Tafel slopes were determined experimentally, but the high resistance of the concrete and consequent change in electropotential presented difficulties in accurate determination of the slopes.

Israel Cornet of the University of California, Berkeley, presented a paper on "Laboratory Testing and Monitoring of Stray Current Corrosion of Prestressed Concrete in Seawater." He described a test in which concrete specimens containing a stressed steel wire and submerged in seawater were subjected to a known impressed current. The specimens were monitored by measuring potential and by visual inspection for up to 83 weeks. A qualitative evaluation of the specimens was made when they were opened at the conclusion of the exposure phase of the test. Where considerable localized corrosion attack was found, there was a "great reduction" in breaking strength for a given ampere-hour exposure level. Where the corrosion was more uniformly distributed, an equal ampere-hour exposure gave "less reduction" in fracture strength. He noted that the corrosion damage did not relate to resistance in the electrochemical circuit and did not correlate with cracks in the concrete—in fact, quantitative prediction of the damage was not achieved by any of the methods of monitoring used. Emphasis was placed on the great degree of caution needed with cathodic protection systems in the vicinity of prestressed concrete structures.

A second paper dealing with prestressed reinforcing was presented by J. C. Griess of Oak Ridge National Laboratory. In "Corrosion of Steel Tendons in Prestressed Concrete Pressure Vessels," Griess reported that ASTM 416 steel is ordinarily protected very well by concrete or grout. Where failures have occurred, poor construction practices have generally been the cause. Stress corrosion cracking, possibly with hydrogen embrittlement, occurred with unprotected steel specimens in neutral or acid H_2S or NH_4NO_3, but not when protected with grout or organic corrosion inhibitors. Oxygen access substantially increased corrosion of unstressed specimens.

"Influence of Selected Chelating Admixtures Upon Concrete Cracking Due to Embedded Metal Corrosion" was presented by William Hartt of Florida Atlantic University. He used a series of admixtures to test the commonly accepted theory that concrete disruption is due to generation of solid products of increasing volume near corrosion sites. Chelating agents (EDTA and related compounds and triethanolamine), which might enhance the solubility of corrosion products, were admixed in concrete for cracking tests using impressed current. Although the chelating agents did increase iron solubility and reduced steel corrosion in simulated concrete pore water, the cracking tests gave failure times of 30 to 150 percent of that for concrete specimens with no admixture, which was judged to be relatively small change. Lower concrete compressive strengths obtained with these chelating compounds as admixtures seem to discourage their commercial use.

The paper given by A. M. Rosenberg of W. R. Grace & Company, "Improved Test Methods for Determining Corrosion Inhibition by Calcium Nitrite in Concrete," represents a continuation of work first presented in *STP 629* on an admixture which is now commercially available and which has been used in actual highway bridges. Simulated bridge decks, 1.8 by 0.6 m by 15.2 cm thick (6 by 2 ft by 6 in.), with different levels of calcium nitrite were salted daily in an accelerated test similar to that used by the FHWA. Collection of open-circuit potentials was materially speeded up by use of a microprocessor. Reduction of the extensive data to corrosion maps of each deck showed the magnitude of inhibition of corrosion in calcium nitrite-treated decks. The inhibitor admixture was shown to be compatible with other admixtures such as air entrainers, retarders, and superplasticizers. Strength of the concrete was improved. Use of two reference electrodes may enable corrosion surveys, with some limitations, without a direct connection to the often difficult-to-access rebar network of a structure.

Robert Heidersbach, University of Rhode Island, reported on "The Degradation of Metal-Fiber-Reinforced Concrete Exposed to a Marine Environment." Metal-fiber-reinforced concrete has application potential in marine environments where added cost may be justified. Metal-fiber-reinforced concrete requires a higher cement content (to coat the fiber surfaces fully) as well as a smaller aggregate size than conventional concrete, and a relatively high water/cement ratio is required. Heidersbach recommended that stainless steel fibers be used to minimize corrosion. Significant corrosion did occur on exposed beams containing carbon steel fibers but not when stainless steel was used. Freeze-thaw deterioration on control specimens was adequately controlled by air entrainment, so the expected benefits of fiber reinforcement were not observed in a comparative test.

Loren Flick presented "Corrosion of Reinforcement in Beams of Internally Sealed Concrete Under Load" from his master's degree thesis under Professor J. P. Lloyd of Oklahoma State University. Concrete may be internally sealed against migration of the aqueous phase by introducing small [0.3 mm (0.012 in.)] wax beads into plastic concrete, then melting the wax after the concrete has been cured. Plain and internally sealed reinforced concrete was subjected to steady and cyclic loads up to 138 MPa (20 ksi) on the test specimens. Flick found that internally sealed concrete was resistant to corrosion under static load. Under cyclic loading, however, cracks developed, acted like pumps, and accelerated the penetration of chlorides, leading to active corrosion potentials. Lime leaching did not appear to hasten corrosion.

Kenneth W. J. Treadaway of the Building Research Establishment, Garston, England, discussed part of a long program to investigate the durability of galvanized steel as concrete reinforcement. Other especially durable steels, such as stainless types 405, 430, 302, 315, and 316, are also being examined. Similar performance, as measured by cracking of cover believed to be induced by expansive corrosion products, has been exhibited by mild steel, high-yield-strength steel, and galvanized steel in good quality chloride-free concrete made from dense aggregate. An increased incidence of cracking for lightweight aggregate and small depths of cover was noted when galvanized bar was not used. Carbonation of the concrete seemed to be a necessary first step before corrosion and proceeded faster in lightweight concrete. Treadaway also made a strong case for adoption of standardized prisms for laboratory tests. The prisms used in the Building Research Establishment programs suggest an approach to such a standard and appear to offer a reasonable starting point for such a development.

The paper by David Stark of the Portland Cement Association on "Measurement Techniques and Evaluation of Galvanized Reinforcing Steel in Concrete Structures in Bermuda," illustrated the performance of galvanized reinforcement in marine environments. This paper presents two innovative analytical techniques for (1) determining actual chloride concentrations at the reinforcing bar and (2) examination of the bar-concrete interface without disturbing the products of corrosion. In the case of galvanized reinforcing steel, these techniques allowed the author to demonstrate that relatively high chloride concentration levels were actually at the bar level. This suggests that the mechanical effects of the products of corrosion are not a negative performance factor for galvanized rebar. In addition, the bar-concrete interface technique permits a relatively accurate determination of the coating thickness on the bar when it was placed. From this, an average corrosion rate may be calculated.

David Corderoy of the School of Metallurgy, University of New South Wales, Australia, performed accelerated tests using impressed current to study the "Passivation of Galvanized Reinforcement by Inhibitor Anions." The placement of concrete around galvanized reinforcing steel may lead to evolution of gas which decreases bond by rendering the adjacent concrete spongy. Bare zinc electrodes treated with sodium chromate in saturated calcium hydroxide solutions passivated (no gas evolution), even in the presence of 0.35 percent chloride ion, but they required four times as much additive if the chromium was present in the form of chromic oxide. The presentation described tests of rebars covered by 10 to 65 mm (0.4 to 2.6 in.) of concrete. In these tests, a linear relationship was observed between depth of cover and first rust staining, with galvanized bars giving approximately 50 percent longer life. Smaller bars tended to give more rapid staining, perhaps because of subsidence cracking.

Cornet's second paper at the symposium, "Critique of Testing Procedures Related to Measuring the Performance of Galvanized Steel Reinforcement in Concrete," was followed by a forum discussion on the many aspects of measuring and evaluating performance of protective systems for reinforced concrete structures subjected to chlorides. Cornet used the literature on galvanized reinforcement to illustrate the care required and difficulties involved in conducting and evaluating corrosion tests for materials. This choice was the apparent consequence of his intimate familiarity and nearly two decades of experience with galvanized reinforcement, coupled with the existence of a fairly significant international body of literature on that subject. He illustrated his concerns about conducting and interpreting tests by showing that factors such as mechanical properties of reinforced concrete structures, as well as test methodologies (that is, corrosion potential measurement, polarization data, and accelerated tests), can create significant barriers to evaluating and extrapolating test results.

Cornet's presentation preceded a wide-ranging discussion which reflected intense interest by many of those present in defining, or at least listing, the parameters which must be investigated further, among which are

1. steel: rebar composition, rolling history, surface condition;
2. sand, stone, and cement: cement factor, cement composition, aggregate permeability, contamination;
3. concrete: nonuniformity, permeability, compaction, cover depth;
4. chloride: determining aqueous concentration at the steel, rather than in surrounding concrete; standardization of analyses of total and water-soluble chloride;
5. corrosion rate: how affected by conductivity and polarization; how related to potential measurements; and
6. structural: relating loss of cross section of reinforcement to structural integrity; relating individual measurements to condition of the full structure.

Overall, the papers again strongly suggest the extreme complexity of the problem and the need for continued applied and basic research effort in this area. It was pointed out repeatedly that collaboration between laboratory investigations and field studies must continue and be expanded. There do not appear to be any simple or readymade solutions to the problems investigated. As a consequence, effort must be intensified if satisfactory answers are to be achieved within a reasonable length of time.

In summary, the international and interdisciplinary opportunity presented to corrosion technologists by the Bal Harbour symposium appeared to benefit authors and participants alike. We hope this symposium has served as a vehicle to expand dialogue among other practitioners and interested parties and will serve as a stimulus to further exchange of data and information on this subject.

D. E. Tonini

American Hot Dip Galvanizers Association, Inc., Washington, D. C. 20005; symposium cochairman and coeditor.

J. M. Gaidis

W. R. Grace & Co., Columbia Md. 21044; symposium cochairman and coeditor.

Index

D

O

P

R

S

T

W

Z